Ask an Astronaut

Tim Peake is a European Space Agency (ESA) astronaut. He finished his 186-day Principia mission working on the International Space Station for Expedition 46/47 when he landed back on Earth on 18 June 2016.

He is also a test pilot and served in the British Army Air Corps. He has over 3000 flying hours and has flown more than 30 different types of helicopter and fixed-wing aircraft.

Tim is a Fellow of a number of UK science, aviation and space-based organisations. He is also a STEM ambassador. In the Queen's Birthday Honours, Tim was made a Companion of the Order of St Michael and St George, an award given to those who have given distinguished service overseas, or, in Tim's case, in space.

Tim is the author of the no. 1 bestseller *Hello, is this Planet Earth?*, his book of photographs taken from the International Space Station, which won the Non-fiction Lifestyle 'Book of the Year' award at the 2017 British Book Awards.

He is married with two sons.

Tim is pleased to announce that royalties received from the book will be donated to The Prince's Trust, for which he is an ambassador.

Praise for ASK AN ASTRONAUT

'Everything you ever wanted to know about life in space' *The Times*

'A delightful adventure of understanding how and why humans journey into space . . . I feel more ready to go into space than I have ever done, though I'm not quite sure I'll get through the training process . . .' Robin Ince, *The Infinite Monkey Cage*

' *Radio 2*

‘An enjoyable read and an excellent insight into the work, life and responsibilities of these highly skilled individuals . . . the perfect gift for anyone with a passion for space’ *Sky at Night Magazine*

‘Peake’s honest and detailed answers combine to give a complete picture of an astronaut’s life . . . charming and informative’
Daily Express

‘Any kid who wants to be an astronaut or an adult who wanted to be – which is basically everyone – will just love it’
Lorraine Kelly

‘Fascinating and informative’ *S Magazine*

Also available by Tim Peake

Hello, is this Planet Earth?

TIM
PEAKE

Ask an Astronaut

arrow books

1 3 5 7 9 10 8 6 4 2

Arrow Books
20 Vauxhall Bridge Road
London SW1V 2SA

Arrow Books is part of the Penguin Random House group of companies
whose addresses can be found at global.penguinrandomhouse.com.

ILLUSTRATIONS BY ED GRACE - WWW.EDGRACE.CO.UK

First published by Century in 2017
First published in paperback by Arrow Books in 2018

www.penguin.co.uk

A CIP catalogue record for this book is available from the British Library.

ISBN 9781784759483

Printed and bound in Great Britain by Clays Ltd, St Ives plc

To my parents,
who gave me the love, support and
encouragement to follow my passions
and seek answers to my questions.

CONTENTS

LIST OF QUESTIONS

Introduction

Launch

Life and Work on the ISS

Earth and Space

Return to Earth

'Live as if you were to die tomorrow. Learn as if you were to live forever.'

Mahatma Gandhi

'The important thing is not to stop questioning. Curiosity has its own reason for existing.'

Albert Einstein

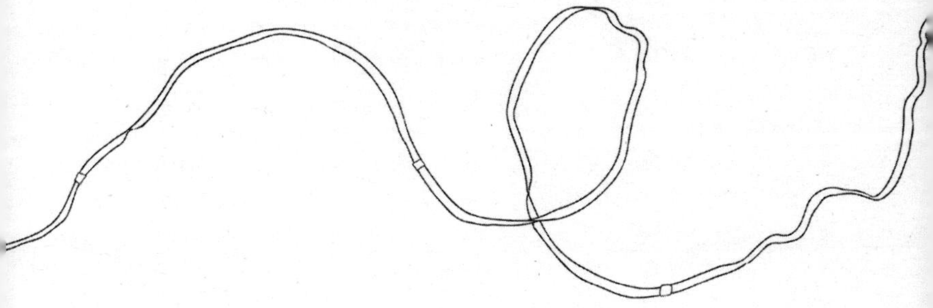

INTRODUCTION

Q *My first question is simple: How can I become an astronaut? – Alexander Timmins, Year 9, Chichester Free School*

A Well, you've picked a brilliant career to pursue, Alexander!

Just as the Apollo missions in the 1960s took a giant leap for mankind, so we are now on the cusp of a new golden age of space exploration. In the coming decades we can expect to colonise the Moon, set foot on Mars and travel deeper into our solar system than ever before. These dreams of human endeavour are now within touching distance, and we can all be a part of this remarkable journey.

You might say this whole book is dedicated to answering your question. And that's because the response is not straightforward, since there is no set path to becoming an astronaut. I was 43 years old when I arrived on board the International Space Station (ISS) on 15 December 2015. I felt enormously privileged to be there and to be following the same career path as men and women I had revered all my life. It was hard to believe I had been fortunate enough to join this exclusive band of spacefarers.

A total of 545 people from 37 different nations had reached Earth's orbit before me, since Yuri Gagarin's first intrepid launch on 12 April 1961. As a small cadre of space explorers, we hail from a wide range of careers and backgrounds – school teachers, pilots, engineers, scientists and doctors – and from every corner of the globe. The one thing we all share is a love of exploration and a passion for human spaceflight.

Of course there are certain skills and characteristics that you need to possess as an astronaut, or acquire during training, and I'm confident that by the end of this book you'll have a clear idea of what is the *right stuff* for today's astronauts. Some of these attributes may surprise you – being good at languages, for instance, is extremely useful. Equally important is what you do before you become an astronaut. It's key to find a career that you're passionate about, and to be as good as you can be in that field. As we shall see, academic requirements only get you so far. It is your drive, your enthusiasm and, above all, your personality and character that will enable you to succeed.

Shortly after landing back on Earth I was asked in a press conference if I had a message for the children from my old school. My own journey had begun in a small village outside Chichester, on the south coast of England. It had taken nearly 18 years in the Army and a career as a test pilot to put me in the right place at the right time to become an astronaut. I replied, 'You're looking at a boy who went to Westbourne Primary School, who left school at the age of 18 with three average A-levels, and I've just got back from a six-month mission to space, so my message is: don't let anybody tell you that you can't do anything you set your heart on.'

Make no mistake, becoming an astronaut is not easy. In fact it has been the single hardest thing I've ever done. But it has also been the most rewarding pursuit by far – full of tremendous experiences that I will treasure for the rest of my life.

★

So what is this book? And what's with all the questions? Well, since returning from the ISS, I've been amazed by the warm response from thousands of people wishing to know more about my mission and what it takes to become an astronaut. I've enjoyed answering intriguing questions on every aspect of my mission, from 'Does space smell?' to

‘Is there gravity in space?’ and ‘What’s the grossest thing about living in space?’. There have been novel questions for me, such as ‘Is there a formal protocol for first contact with aliens?’; and more sobering ones, such as ‘What would happen if you were hit by space debris during a spacewalk?’. And of course there are fun ones, such as ‘Can you have a cup of tea in space?’ (the answer is, thankfully, yes!) and ‘How do you go to the toilet in space?’, which is by far the most popular question I get asked, particularly by younger children.

I wanted to answer and expand upon as many of these questions as I could, in order to provide my own definitive account of what being an astronaut is really like – the personal, the profound; the adventure, the astrophysics; the fear, the fun. I hope both the science and the everyday details of life in space are entertaining and informative, and may provide a useful reference or handbook for the next generation of spacefarers. After all, the first person to walk on Mars may well be reading this book.

Using the hashtag #askanastronaut, the project was opened up to users on social media. Many of the wonderful submissions from Twitter and Facebook are included in the book, as indicated by the names after some of the questions. In other cases, where more than one person asked the same or a similar question, I have created an amalgamation. Thank you so much to everyone who contributed to the project. Even if your name is not included on these pages, your curiosity and thoughtfulness have played an enormous role in shaping the book, and I am hugely grateful for your inquisitiveness.

I have tried to cover all of the key parts of my mission in this book, which is structured into seven chapters: ‘Launch’, ‘Training’, ‘Life and Work on the ISS’, ‘Spacewalking’, ‘Earth and Space’, ‘Return to Earth’ and ‘Looking to the Future’. As well as answering your questions, I have also answered some of my own. I have tried to share insights into my journey to space, from describing the training and preparation, to the science behind the ISS, the experiments on board, the beauty of Earth from 400 km up, the thrill of travelling at supersonic speeds through the atmosphere, the excitement and perils of walking in space, the camaraderie of the crew, and the change of perspective as a result of these astonishing experiences.

Writing and researching the book, and reliving my time on the space station, has been a fascinating experience. I hope that with such a variety of subjects covered, it will be of interest to readers of all ages. Some of the answers are quite long and technical, while others are much shorter. So to give you a taste of what is to come, here are a few quick questions and answers to get you started.

Q *If you see 16 sunrises a day in space while orbiting Earth, when do astronauts celebrate New Year?*

A Since the time zone on the space station is the equivalent of Greenwich Mean Time (GMT), the clock strikes midnight on New Year at the same time as in London. If nothing else, we need more British astronauts in space purely for this reason! However, each astronaut on board will usually celebrate New Year when it strikes midnight in their home nation.

Q *Did you miss the weather on Earth when you were in space, and what did you miss the most?*

A This is going to sound strange, but I really missed the rain. I had no access to a shower for six months and I love exercising outdoors, so when I was running on a treadmill, confined to a warm module on the space station, the idea of cold drizzle on my face sounded blissful.

Q *What was your luxury item on board?*

A The item that I got the most pleasure from was definitely my camera, since photography became a new-found passion whilst I was in space and a source of excitement, wonder and satisfaction. I treasure the photographs I took from space and, even looking back through them now, I can always remember exactly when and where the space station was when I took them. However, I wouldn't describe our cameras as luxury items, since we regularly used them for valuable Earth observation science. In terms of pure decadence, I think the best luxury item was a small coolbox that arrived on the Dragon resupply spacecraft, addressed to the crew from the kind folk at SpaceX (which manufactures rockets and spacecraft), which just happened to be packed full of ice cream!

Q *In the build-up to your mission, did you become less afraid of going into space, the more knowledge you acquired?*

A During astronaut training (which we will explore in detail in Chapter 2), as you gain knowledge, it certainly helps to allay some of the concerns you may have about the higher-risk parts of a mission, such as performing a spacewalk, launch, re-entry and emergency situations. More importantly, knowledge gives you the ability to generate options to deal with difficult situations, and prevents you from making poor choices in the first place. As NASA astronaut and commander of Apollo 8, Frank Borman, once quipped, 'A superior pilot uses his superior judgment to avoid situations which require the use of his superior skill.'

Our training is exemplary, and all astronauts owe a huge debt to the incredible team of trainers and instructors who dedicate themselves to ensuring that we are fully prepared to execute our mission safely and effectively.

I walked out to the launch pad feeling completely ready to go to space, eagerly anticipating the thrill and excitement of the best ride of my life. If you had asked me right then if I was afraid of going to space, my immediate reaction would have been to say, 'No way!' However, flying to space involves risks that no amount of knowledge, training or preparation can mitigate. All astronauts understand and weigh up these risks prior to launch, but no one can guarantee that something catastrophic (by which I mean loss of the spacecraft or crew) will not occur. Saying goodbye to my family just before launch was the hardest thing I have ever had to do. By strapping into a rocket you are voluntarily rolling the dice, and there's a chance you will not be coming home.

Fear is a feeling induced by perceived danger, and if someone doesn't perceive a danger when sitting on top of ten storeys of highly flammable rocket fuel, then they probably don't fully understand their predicament! A more accurate answer would have been to say, 'Sure, there's a part of me that's afraid, but I've dealt with that and it's not what I'm thinking about right now.'

This seems like a good time to introduce the first chapter: LAUNCH!

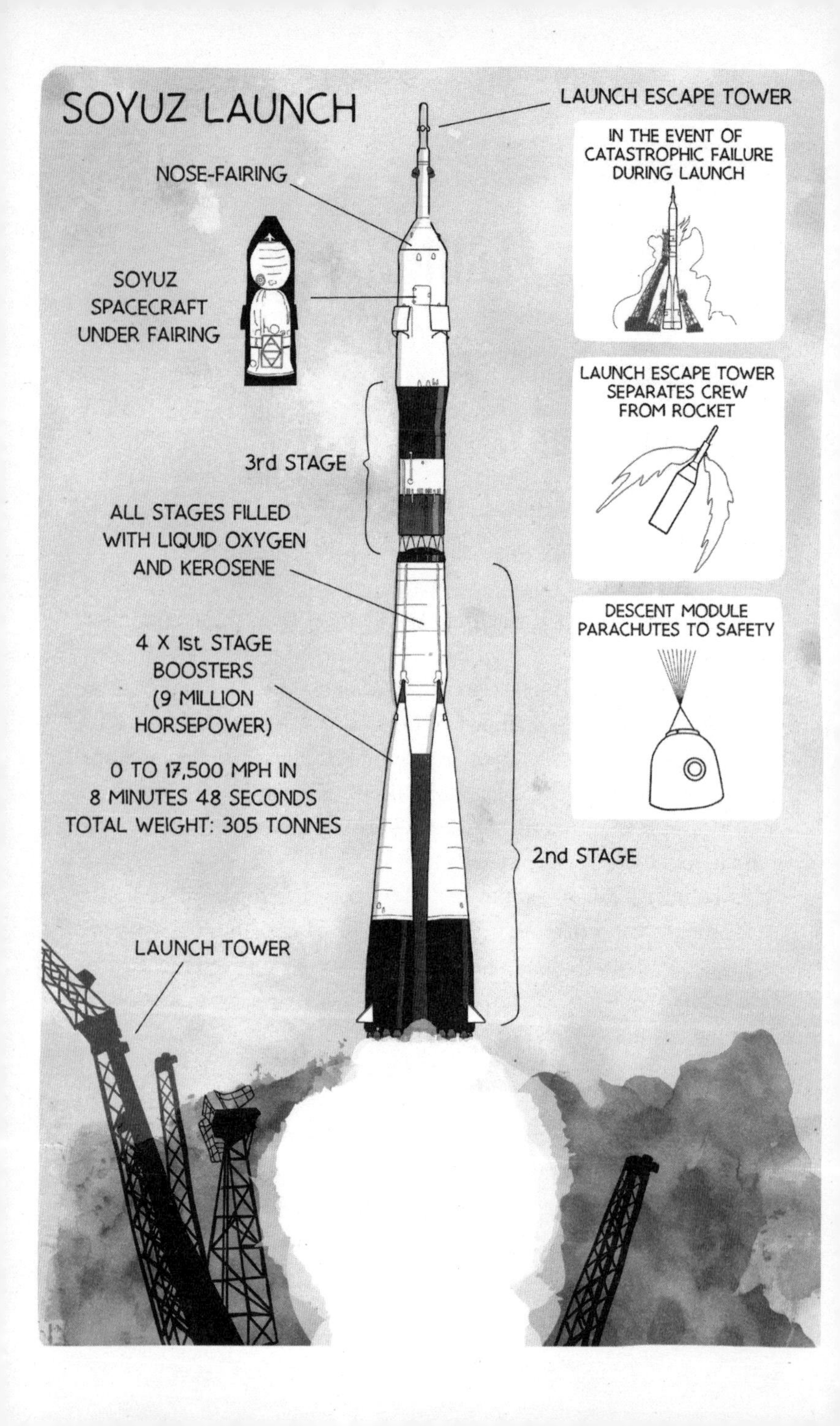
SOYUZ LAUNCH
LAUNCH ESCAPE TOWER
IN THE EVENT OF CATASTROPHIC FAILURE DURING LAUNCH
NOSE-FAIRING
SOYUZ SPACECRAFT UNDER FAIRING
LAUNCH ESCAPE TOWER SEPARATES CREW FROM ROCKET
3rd STAGE
ALL STAGES FILLED WITH LIQUID OXYGEN AND KEROSENE
DESCENT MODULE PARACHUTES TO SAFETY
4 X 1st STAGE BOOSTERS (9 MILLION HORSEPOWER)
0 TO 17,500 MPH IN 8 MINUTES 48 SECONDS
TOTAL WEIGHT: 305 TONNES
2nd STAGE
LAUNCH TOWER

LAUNCH

Q *What does it feel like to sit on top of a 300-tonne rocket?*

A 15 December 2015, Kazakhstan. 14.33 local time. Launch minus 2 hours, 30 minutes.

I was standing 50 metres above the launch site, at the top of the glistening Soyuz rocket, waiting to climb inside. It was a gloriously clear winter's day. Looking out over the sprawling Baikonur Cosmodrome and the vast expanse of grassland that was the Kazakh Steppe, my senses were in overdrive absorbing the last sights, smells and sounds of planet Earth before I left for six months.

As I climbed aboard our tiny capsule, situated within the nose-fairing of the rocket, the vehicle felt completely alive beneath me. Cryogenic fuel was continuously boiling off, covering the base of the rocket in an eerie white fog. This sub-zero propellant caused a layer of thin ice to cover the lower two-thirds of the rocket, transforming the usual orange-and-green livery of the Soyuz into a dazzling white in the afternoon sunshine. We had enjoyed a close-up view of the rocket as we took the lift-ride up to our capsule. With it fully fuelled with 300 tonnes of liquid oxygen and kerosene, hissing and steaming within its metal support structure that held it in place prior to ignition, you get a real sense of the incredible engineering it takes to escape the force of Earth's gravity.

I've strapped into many aircraft in my career, but I'm certain nothing will ever come close to the exhilaration of climbing aboard a rocket prior to launch. I didn't feel nervous; quite the opposite. I had waited a long time for this moment and, despite trying to maintain a calm, professional focus, I was only too aware of a boyish excitement building deep within me.

We always climb into the capsule in a specific order. The first one in is the left-seater (Tim Kopra in our case), then the right-seater (myself), then finally the Soyuz commander (Yuri Malenchenko). First, we had to enter the crammed habitation module through a horizontal hatch and then wiggle our way, feet first, down through a vertical hatch to enter the descent module. There's no ladder, but there are footholds that help.

We had to be very careful squeezing past the vertical hatch as it contained the antenna, which would be needed six months later to transmit our location to the search-and-rescue crews after landing. It was a real squeeze getting into the seat. Unlike the Soyuz simulator back in Star City, Russia, where we had trained, the spacecraft was packed to full capacity with cargo. Initially I dropped down into the commander's seat and then cautiously shifted across, feet first, into my right-hand seat. Everything had to be done very slowly and carefully. This was not the time to tear my spacesuit or cause damage to the spacecraft. I thought of all the times I'd been caving, during my training, and was grateful for having had some experience of working in extremely confined spaces.

As soon as I was in my seat, there were two electrical cables and two hoses that had to be connected to the Sokol spacesuit. The electrical cables were for my communications headset and medical harness, which I had donned earlier. All crew wear a medical harness next to their chest, which measures heart rate and breathing rate, with the data being transmitted back to our flight surgeons. The two hoses were for air (for cooling and ventilation) and 100 per cent oxygen (used only in the case of an emergency depressurisation). Having made these connections, the next steps were to connect my knee braces, which would prevent injury to my legs during any high g-loading that might occur during launch, and to secure my five-point harness. There was just enough room for one ground-crew member to help me strap in and hand me my checklists.

As I counted the minutes until launch, meticulously reviewing the checklist one last time and mentally visualising the crucial minutes and hours ahead, there was time for one final tradition to be observed, to get the adrenaline flowing. Each cosmonaut is allowed three songs to be piped into the capsule before lift-off. I had elected for 'Don't Stop Me Now' by Queen, 'Beautiful Day' by U2 and 'A Sky Full of Stars' by Coldplay. As the crew's chosen compilation faded, and with only moments to go until ignition, there was one last surprise. Through our headsets, and drowning out the loud burr of the rocket, we heard the familiar synthesiser notes and guitar chords of 'The Final Countdown' by Europe, chosen by our Soyuz instructor – who says the Russians don't have a sense of humour!

★

The first time I watched a Soyuz rocket launch (other than on a screen) was in June 2015, six months prior to my own launch. Along with fellow Soyuz crewmates Yuri Malenchenko and Tim Kopra, I had travelled to Baikonur in Kazakhstan as backup for the Expedition 44/45 crew (the previous team of astronauts to visit the ISS). Our job was to mirror the prime crew and support them in any way we could. Although we were the backup crew and were ready to launch to space, having passed all the necessary exams a couple of weeks earlier, the likelihood of us actually replacing the prime crew was very slim. However, being in Baikonur gave me the perfect opportunity to see a full dress rehearsal, including watching my first rocket launch. I had tried to watch the Space Shuttle *Discovery* launch several years earlier, when fellow European Space Agency (ESA) astronaut Christer Fuglesang launched from Kennedy Space Center in Florida. But the first launch was delayed due to weather concerns, the second attempt was called off due to an anomaly in one of the Orbiter's fuel valves, and days later, just as *Discovery* launched into space, I was on a plane heading back to Europe to begin my training at the European Astronaut Centre in Germany . . . Murphy's law!

Watching the Soyuz launch in June 2015 more than made up for any earlier disappointment. What made it more spectacular was how close we were when it lifted off the launch pad. Yuri, Tim and I were sitting on the roof of the search-and-rescue tower, about 1.5 km away from the rocket. It was gone 3 a.m. on a beautifully clear night and, when I saw the main engines light up, followed by a deep roar a few seconds later, a huge grin spread across my face. But my expression soon changed to one of astonishment. What I had heard so far was merely the engines at intermediate thrust, when there's a brief pause for a checkout. As the main engines opened up to full power, the noise engulfed me – a powerful rumble of deep bass notes that reverberated around my chest cavity. Just when I thought it couldn't be any more impressive, the Soyuz lifted off the launch pad and, as it climbed, a deafening crackle filled the air.

A few months later, sitting in my Soyuz seat at just past 5 p.m. local time, I was listening intently to our instructor's voice through my headset, whilst my eyes were glued to the digital clock in front of me. Of all the times in your life when you expect a good old-fashioned countdown, a rocket launch would be one of them. Disappointingly, this is not the case! As the engines ignited to intermediate thrust and the turbopumps accelerated to flight speed, our instructor was announcing the stages of this final sequence, giving the crew cues as to when launch would occur, but there is no actual countdown. As we heard the call that the engines were at full thrust, which occurs five seconds prior to launch, the feeling of pent-up power from the rocket beneath us was immense. In the final seconds before lift-off, the noise and vibration inside the capsule are such that you have no idea whether you have left the launch pad or not. I felt the rocket sway markedly and noted the clock running over time. We were off! As I heard the distinctive crackle from the brute force of the rocket engines and the acceleration started to kick in, I reflected back to six months ago and how it must feel to everyone watching.

Oddly, the noise is not as impressive inside the capsule as it is outside. Don't get me wrong: it's still extremely loud. However, when you're wearing a communications cap under a sealed spacesuit helmet, it provides a fair degree of sound insulation. What's far more impressive inside the spacecraft are the sheer energy, vibration and acceleration that you experience; it feels almost visceral. But there's no violent explosion, no ringing of ears, and you can't see anything out of the windows, since the rocket's nose-fairing is still protecting the spacecraft at this point.

In a matter of a few minutes we would be travelling at speeds of 8 km per second – the equivalent of London to Edinburgh in under 90 seconds. It was hard to contain the thrill; I couldn't stop smiling.

★

This chapter chronicles the launch of the Soyuz rocket from the moment of ignition to the moment it docks with the ISS. Flying to space has to be one of the most amazing and surreal experiences, but flying to space with the Russians, as we did, is perhaps even more remarkable. The Russian philosophy of 'If it ain't broke, don't fix it' applies not only in their approach to engineering, but indeed to everything surrounding human spaceflight, which is steeped in history and tradition. If it worked for Yuri Gagarin, then it will work for the cosmonauts and astronauts of today. This means that the weeks, days and hours before launch are filled not only with essential operational tasks, but also with many important traditions and rituals that have to be upheld. We'll pick up the specifics of launch day again a few pages later on, but first let's examine the launch site in a bit more detail.

Q *Why do astronauts launch from Kazakhstan?*

A The Baikonur Cosmodrome, situated in the desert steppe of southern Kazakhstan, is the world's first and largest operational space-launch facility. Ever since the American Space Shuttle programme ended in 2011, it has been the only launch site in the world used to ferry crew to the International Space Station. But this fabled Russian launch pad dates back to the 1950s, when it was built by the Soviets. The first manned spacecraft in human history, Vostok 1, was launched from Baikonur in 1961; and the world's first satellite, Sputnik 1, even earlier in 1957. What makes launches at the Cosmodrome particularly dramatic are the visual pyrotechnics. Unlike at some other launch sites around the world, where water is hosed under the rockets at ignition to douse the flames and muffle the sound, at Baikonur they don't use water, partly because of the desert setting. This makes for a fiery lift-off!

As you would expect, a lot of thought and planning goes into the location of such a launch site. If you want to optimise the efficiency of getting cargo into space, then you can actually use Earth's west-to-east rotation to give you a bit of a free kick in that direction. This 'free' velocity is not insignificant and is at its maximum at the equator, around 1,670 km/h . . . that's faster than the speed of sound! Of course when you stand at the equator you don't feel this velocity because the air around you is travelling at the same speed. But when you're launching into space, that extra punch really matters. As you travel away from the equator, the free velocity decreases, until it reaches zero at the north and south poles – the point at which the surface of Earth is just spinning on its axis of rotation.

So a rocket launching nearer the equator has a head start, which means less fuel is required to get it to orbit, which in turn means that a heavier payload can be carried, in place of fuel. But take a look at a world atlas and you'll see that Russia is not blessed with low latitudes. The vast majority of the territory is above 50°N and, having spent a few winters in Russia, I can vouch for its non-tropical climate.

Baikonur, in Kazakhstan, sits at 46°N. Not exactly equatorial, I hear you cry, but at least it's farther south than most of Russia. Of course there's more to the story than simply latitude. The site was originally selected in 1955, as a test range for the world's first intercontinental ballistic missile. Only later was it expanded to include launch facilities for space flights. As a missile test range, it had to be surrounded by flat plains in order to ensure uninterrupted radio signals from the ground-control stations. Additionally, the missile trajectory – headed for test targets 7,000 km away in Kamchatka – had to be away from populated areas. Baikonur and the Kazakh Steppe fulfilled all these criteria, in addition to having a water supply from the Syr Darya river and the Moscow–Tashkent railroad not a million miles away.

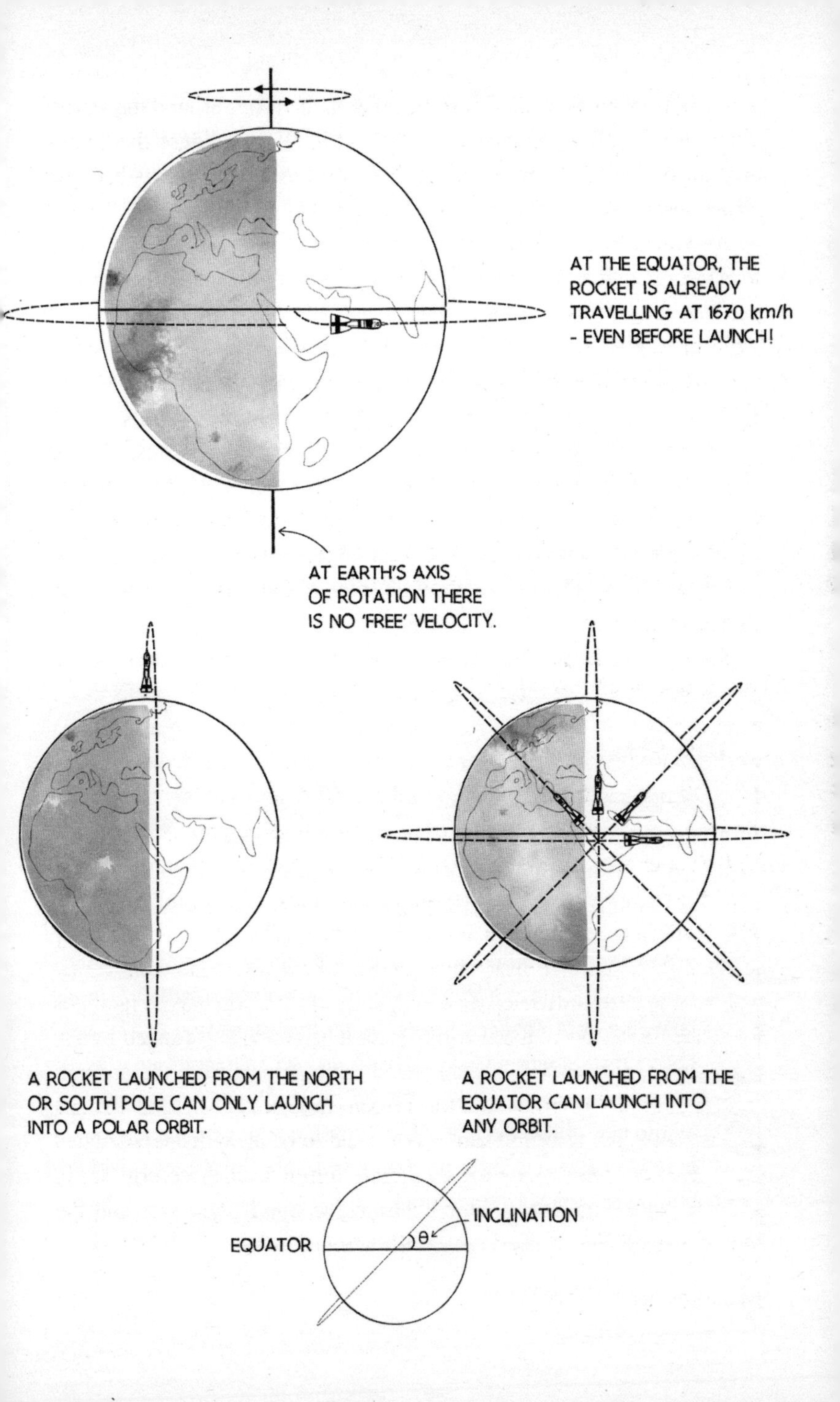
AT THE EQUATOR, THE ROCKET IS ALREADY TRAVELLING AT 1670 km/h - EVEN BEFORE LAUNCH!
AT EARTH'S AXIS OF ROTATION THERE IS NO 'FREE' VELOCITY.
A ROCKET LAUNCHED FROM THE NORTH OR SOUTH POLE CAN ONLY LAUNCH INTO A POLAR ORBIT.
A ROCKET LAUNCHED FROM THE EQUATOR CAN LAUNCH INTO ANY ORBIT.
INCLINATION
θ
EQUATOR

There's another reason for launching as close to the equator as possible, in addition to cashing in on Earth's 'free' rotational velocity. It gives you a greater choice of orbital inclination. This measures the tilt of an orbit and is expressed as an angle between the equator and the axis of direction of the orbiting object.

An easy way to imagine this is to think of a rocket launched from the north pole. It doesn't matter which direction you point that rocket, it can only go south. It will be inserted into a polar orbit (passing over the north and south poles of the planet), with an inclination of 90 degrees. Conversely, a rocket launched from the equator can point in any direction and can be inserted into any orbital inclination. For locations in between, the choice of orbital inclination is limited by the launch site's latitude.

This rule can be broken by throwing fuel at the problem and doing an 'orbital plane-change manoeuvre'. However, it takes an awful lot of fuel to tilt an object's inclination, once inserted into orbit, and most mission planners try to avoid this at all costs.

Did you know?

- As Baikonur developed into the world's premier space launch facility, its latitude later became a major constraint when deciding the orbital inclination of the International Space Station, which is 51.6 degrees.
- There are many other launch sites around the world. The United States have a long history of human spaceflight launches from Kennedy Space Center in Florida, and two new spacecraft (Boeing's CST-100 and SpaceX's Dragon) are due to begin launching crews to the ISS from US soil once again, by the end of this decade. The Chinese use the Jiuquan Satellite Launch Centre, located in the Gobi Desert, for their human spaceflight programme.

Q *How long do astronauts spend in quarantine before launch, and can anyone visit them?*

A The purpose of astronaut quarantine before a mission is to ensure that the prime crew remain fit and healthy, arriving at the ISS free from virus or infection. The length of time spent sequestered like this varies, but it's usually around two weeks. As members of the Expedition 46/47 crew, we spent 15 days in quarantine, and this was a chance to complete some last-minute administration and final Soyuz training prior to launch. By this late stage there wasn't much preparation left to do, so we also had the chance to relax and see family and friends who had come to watch the launch.

Being confined to quarantine didn't mean that we couldn't see anyone, but a strict regime was enforced by the Russian medical staff, which limited face-to-face visits to just a few immediate family members. These members all had to undergo a quick medical examination from the flight surgeon prior to each visit with a crew member. Unsurprisingly, this luxury wasn't extended to youngsters under the age of 12, who have a tendency to be walking biohazards – especially during the winter months. Baikonur is rarely above freezing in December, which made it tough on my two young boys, who sometimes struggled to understand why they could only see their daddy sitting behind a big glass panel.

However, quarantine is most definitely a necessary precaution, and I sympathise with the medical staff who have to enforce it. This lesson was reinforced as early as 1968, during the 11-day Apollo 7 mission. First Wally Schirra, the veteran commander, and then rookie crewmates Walt Cunningham and Donn Eisele all developed serious head-colds. That's the problem with a spacecraft – the confined space, recirculated air, reduced ability to wash and frequent contact with common surfaces and other crew make for a viral Mardi Gras, and soon everyone on board has succumbed to the pathogen. We go to great lengths to ensure that the space station remains clean, sanitised and as healthy a place to live and work as possible – and that task starts in quarantine, prior to launch.

Q *What do you do to prepare, on the day of launch?*

A Unsurprisingly, on launch day everything is determined by the time of the launch. Although the exact time of each Soyuz launch will vary, the strict schedule leading up to launch is set in stone. Every Soyuz crew goes through the same precise routine – everything starts on time, everything stops on time. It's an incredibly thorough process, which is executed with the utmost efficiency, with plenty of buffers built in to ensure there is never any rush. This not only prevents anything important being forgotten, but the whole process ensures that the crew is delivered, dressed and ready for the bus ride to the launch pad, in the correct frame of mind – confident, relaxed and raring to go. Here is the timeline of morning activities for our launch:

07.55–08.05	Wake up, hygiene procedures (10 mins)
08.05–08.15	Medical check (10 mins)
08.15–09.15	Special medical procedures (60 mins)
09.15–09.35	Hygiene wash (shower) (20 mins)
09.35–09.40	Microbiological control (5 mins)
09.40–09.50	Special medication of the skin (10 mins)
09.50–09.55	Putting on underwear to go under the Sokol spacesuit (5 mins)
09.55–10.05	Walk to the cosmonaut hotel (10 mins)
10.05–10.35	Meal, toilet (30 mins)
10.35–10.55	Cosmonaut farewell (20 mins)
10.55–11.00	Traditional signing of the crew-room door (5 mins)
11.00–11.05	Religious custom (5 mins)
11.05–11.10	Getting on the buses (5 mins)
11.10	Departure to Building 254 (to don Sokol spacesuit)

The ten-minute medical check was the same one we had been having each morning during our time in quarantine: basic vital signs, plus weight check. This was to try and ensure that we had not picked up a virus or infection and, of equal importance, that we had not put on (too much!) weight. The food was so good and plentiful in the days leading up to launch that this was not as easy as it sounds. Changes in crew weight affect a spacecraft's centre of mass, which had been precisely calculated to ensure a safe and accurate journey into space. Fortunately, I managed to maintain my 70 kg . . . more or less.

I should take a minute to describe the 'special medical procedures'. It will not come as a surprise to many to learn that most astronauts wear adult incontinence pants during launch (otherwise known as nappies, diapers or – as NASA likes to call them – 'Maximum Absorbency Garments' or MAGs). This has nothing to do with protecting against the excitement of being launched into space on top of a 300-tonne firework, but instead concerns the simple fact that you are in your spacesuit for about ten hours on launch day – and that's a long time for anyone to hold on, even those with the strongest of bladders!

However, the 'special medical procedures' referred to on the timeline were not related to bodily function No. 1. They were, in fact, all about No. 2. As a means of avoiding any unwanted distractions on launch day, and to give the digestive system a day or two to get used to microgravity before a call of nature, astronauts are offered an enema before they fly. In fact I was offered the choice of a US-style or Russian-style enema (I can only guess the European, Japanese and Canadian space agencies have yet to perfect their own styles!). For the life of me, I can't remember what the difference was, and by that point I didn't have much spare brain capacity to allocate to the decision anyway – all I can say is that the Russian style worked just fine.

Having been cleansed thoroughly internally, the next stage was for an equally thorough external cleansing. Following a shower using special antimicrobial soap and a dry-off with a sterile towel, our job was to remain stoically naked and call for the flight surgeon. You learn very early on, as a wannabe astronaut, to hang up your pride. The pathway to space is littered with indignities, such as sigmoidoscopies, endoscopies, enemas and no end of prodding and poking. To that end, a quick body wipe-down from the flight surgeon with special antibacterial towels was not hard to endure, and following this we were able to get dressed in our sterile white long johns and long-sleeved underwear.

The last meal before spaceflight is usually shared with the backup crew and the Russian flight surgeon. This was a great opportunity to relax and enjoy some good-hearted banter before things got a bit more serious. Traditionally this is a breakfast menu, with a choice of eggs, bacon, *kasha* (hot cereal, made with grains boiled in milk), bread, ham, cheese, jam or some fruit. There is always plenty of good Russian tea, too. I ate well, knowing that my next hot meal would not be for several hours – and there would be nothing fresh about it. Then, after breakfast, the formalities began. As prime crew, we met up with our partners in a small private room, along with the backup crew and senior management from the various space agencies that were represented. There was an opportunity for toasts to the success of the mission and to the well-being of family and friends left behind (I should add that the prime crew have to stick to water – not champagne or vodka, more's the pity!). Then we said a quick farewell to our partners: the last chance for a few private words before stepping out in front of the cameras.

The first of several traditions on launch day was for each crew member to sign the door where they were staying in the 'Cosmonaut Hotel' in Baikonur. This was a very special moment – to be able to add my signature alongside so many inspirational men and women who had trodden this path before me. Next was the blessing of the crew by a Russian Orthodox priest, waiting at the end of the corridor. Following the blessing, we descended the three flights of stairs leading to the hotel foyer to a catchy Russian rock anthem by Земляне (Earthlings) played at full volume. Called 'Трава у Дома' ('The Grass Beside our Home'), it tells the story of a cosmonaut's love for Earth. Admittedly this was one tradition that didn't date back to 1961, but was a welcome addition a few years later – you certainly felt pretty pumped about jumping on a rocket by the time you got to the bottom of the stairs! Waiting outside the hotel were friends and family to wave us off as we boarded the bus for the 30-minute ride to Building 254, which is where we donned our Sokol spacesuits. There was time for one final farewell to immediate family members from behind the glass, prior to walking out for the bus ride to the launch pad.

Did you know?

Here are some quick-fire facts about the Sokol spacesuit:

- It was introduced in 1973 and was designed for wear inside a spacecraft, not for spacewalking.
- It inflates using 100 per cent oxygen, to protect the crew in the event that the spacecraft loses pressure (there are two spacesuit pressure settings: 0.4 bar or 0.27 bar, for not more than five minutes).
- It is individually tailored to each astronaut.
- It can be self-donned in two to three minutes, although during suit-up on launch day the engineers will usually take around ten minutes, to check that everything is perfect.
- It has a rubberised neck seal, in the event of a water landing.
- It weighs only 10 kg.
- The pressure seal is made by wrapping two rubber bands around the main opening.
- It's pretty comfortable to wear sitting down, but not so much standing up. That's why astronauts always appear hunched over when they walk out to the bus that takes them to the launch pad.

Q *Is it true that astronauts pee on the bus tyre, prior to launch?*

A One of the many wonderful (and sometimes weird) traditions that the Russians follow, prior to launch, is to have a pee en route to the launch pad. Actually, if you are about to be stuck in a rocket for several hours, this makes plain good sense. Tradition has it that Yuri Gagarin was on his way to the launch pad in 1961 and needed to urinate one last time. Little did he realise, when he chose the back-right tyre of the bus for his bathroom break, that he would be setting in stone a ritual that has lasted for well over 50 years.

The only problem is that by this stage the crew is pretty much ready to go to space, fully dressed in a spacesuit that has already been made airtight and leak-checked. As the bus pulled up for the mandatory loo stop, I remember fumbling with the shoelace-type fasteners and rubber-banded pressure seals, undoing the good work that those technicians, with their protective masks and sterile gloves, had painstakingly done for us less than an hour earlier.

However, I was thankful for the opportunity to relieve myself one last time, and the whole experience was made more poignant by the fact that we were also headed to the same launch pad that Gagarin left Earth from, on 12 April 1961.

Did you know?

Other Russian traditions for the prime crew, prior to launch, include:

- A visit to Red Square to lay flowers at the graves of Yuri Gagarin and Sergey Korolev (considered the 'father' of Russia's space programme).
- A breakfast ceremony prior to leaving Star City for Baikonur (in accordance with Russian superstition, everyone sits in silence for a few moments before leaving).
- Planting a tree in the Avenue of the Cosmonauts grove in Baikonur.
- Not watching the roll-out of the Soyuz rocket – that's considered bad luck for the prime crew.
- Having the train that pulls the Soyuz crush coins on the rails, to invite good luck.
- Getting a haircut two days before launch.
- Watching the 1969 film *White Sun of the Desert* on the night before launch.
- Blessing of the Soyuz rocket by a Russian Orthodox priest.
- The Soyuz commander choosing the mascot – usually a small cuddly toy that hangs from the instrument panel and is the first object to float, on reaching orbit.

Q *How did you all fit in that Soyuz capsule?*

A Ha! There's no doubt about it, the Soyuz is a tight squeeze – and that's coming from someone who is 5 feet 8 inches and weighs 70 kg.

It can also be painful sometimes, because you spend a long time in a kind of foetal position, with your knees bent more than 90 degrees, which is not comfortable. However, it's a small price to pay for the ride of your life! Once you are in orbit, you can loosen your harness and float up a little bit out of your seat. It doesn't sound like much, but I found that by doing that, life became a whole lot more comfortable.

The Soyuz descent module is slightly smaller than the Apollo command module (and way smaller than the Space Shuttle or the new Orion deep-space exploration vehicle). Despite being pretty cramped, we had spent so much time training in the simulator that the Soyuz had become a home-from-home. If anything, it felt kind of cosy, and the small space didn't bother me at all. Having said that, I would have the luxury of a short journey to the space station, arriving in a matter of hours after launch. Some crews have to spend two days in that confined space, prior to docking with the ISS.

Q *How much computing power does the Soyuz have?*

A Our Soyuz was a version called TMA-M (Transport Modified Anthropometric), which first flew in October 2010. It replaced 36 obsolete pieces of equipment with 19 modifications from the previous spacecraft, such as seat changes, glass cockpit displays, parachute system, soft-landing jets and three-axis accelerometers. One of the main upgrades was to replace the Argon digital computer that weighed 70 kg (that's not a misprint – 70 kg!). The Argon was a reliable computer that had been used in the Soyuz for more than 30 years. However, its performance statistics were not that impressive, being on a par with the so-called Apollo Guidance Computer used for the Moon landings. The new computer, called a ЦВМ 101 (central computer), whilst being several orders of magnitude better than the old Argon, still pales in comparison to the computing power of an average smartphone. Let's take a look:

	Argon-16	ЦВМ 101	iPhone 7
Processor speed	Up to 200 kHz	6 MHz	2.34 GHz
RAM	3 × 2 kbytes	2 MB	2 GB
ROM	3 × 16 kbytes	2 MB	256 GB
Weight	70 kg	8.3 kg	138 grams

However, the Soyuz does have more than one computer. There's also a TBM (terminal computer) and a КЦП (central post computer), but neither is going to change the fact that the Soyuz really doesn't need a lot of computing power to fly to space!

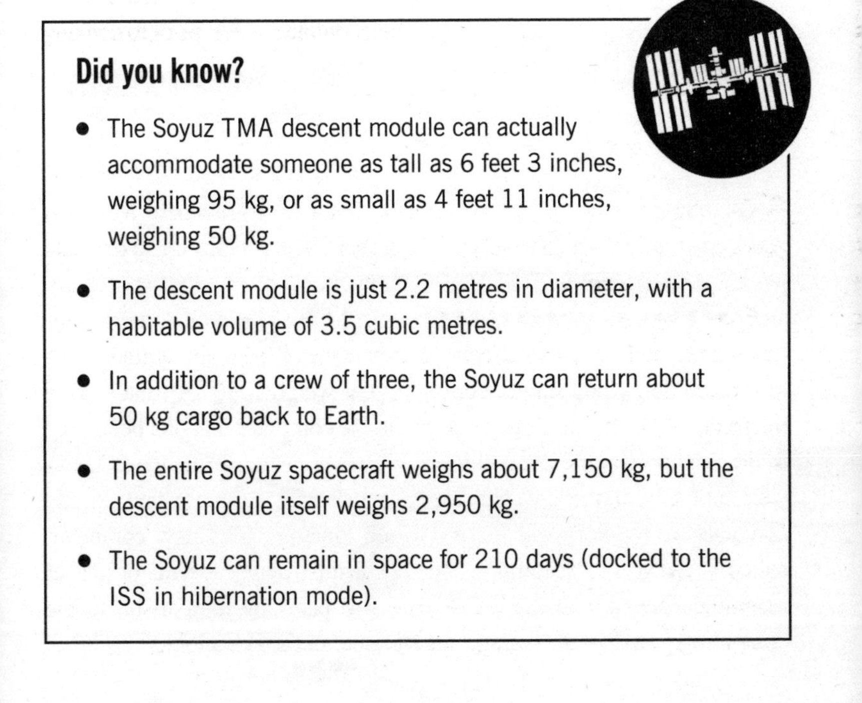

Did you know?

- The Soyuz TMA descent module can actually accommodate someone as tall as 6 feet 3 inches, weighing 95 kg, or as small as 4 feet 11 inches, weighing 50 kg.
- The descent module is just 2.2 metres in diameter, with a habitable volume of 3.5 cubic metres.
- In addition to a crew of three, the Soyuz can return about 50 kg cargo back to Earth.
- The entire Soyuz spacecraft weighs about 7,150 kg, but the descent module itself weighs 2,950 kg.
- The Soyuz can remain in space for 210 days (docked to the ISS in hibernation mode).

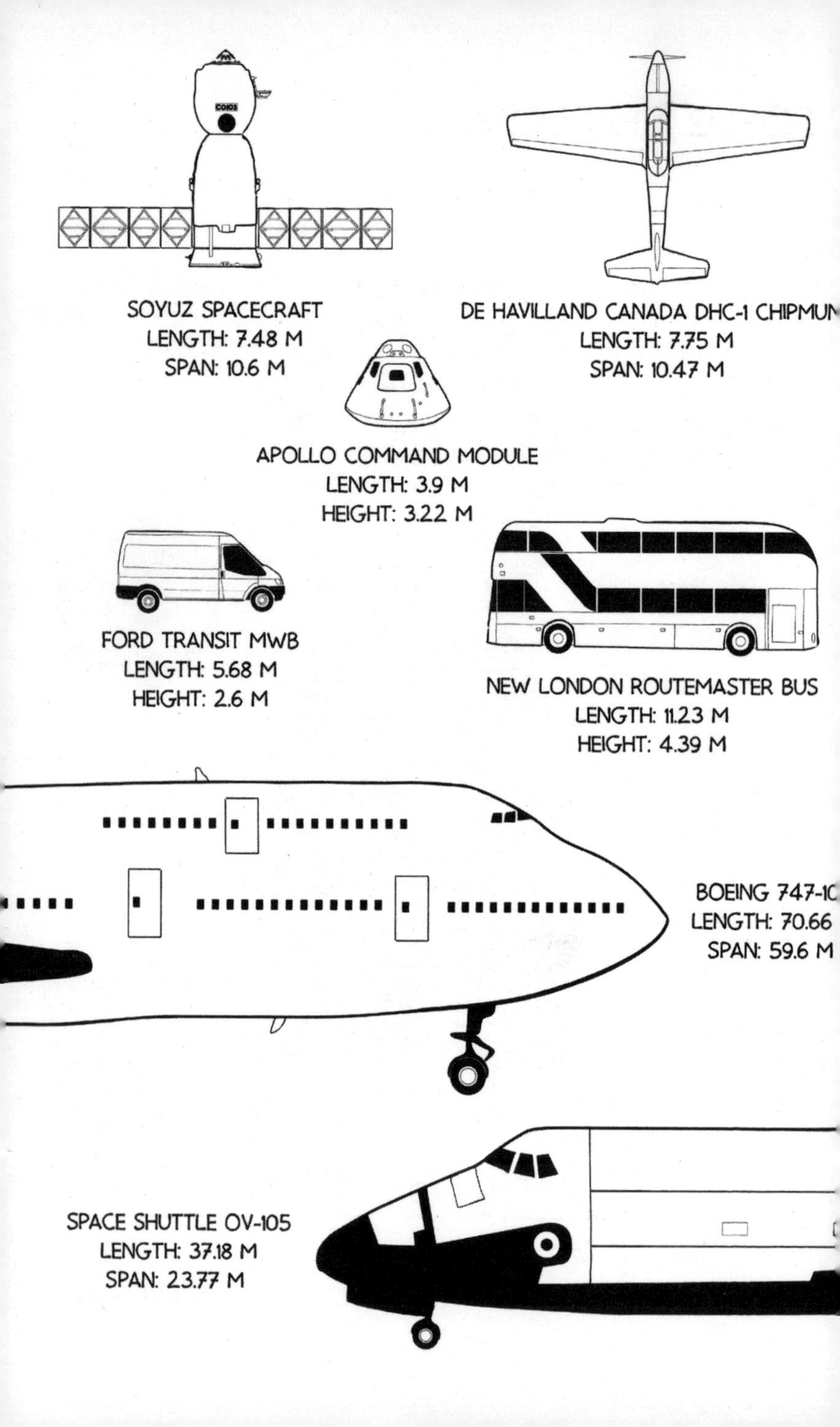
SOYUZ SPACECRAFT
LENGTH: 7.48 M
SPAN: 10.6 M
DE HAVILLAND CANADA DHC-1 CHIPMUN
LENGTH: 7.75 M
SPAN: 10.47 M
APOLLO COMMAND MODULE
LENGTH: 3.9 M
HEIGHT: 3.22 M
FORD TRANSIT MWB
LENGTH: 5.68 M
HEIGHT: 2.6 M
NEW LONDON ROUTEMASTER BUS
LENGTH: 11.23 M
HEIGHT: 4.39 M
BOEING 747-10
LENGTH: 70.66
SPAN: 59.6 M
SPACE SHUTTLE OV-105
LENGTH: 37.18 M
SPAN: 23.77 M

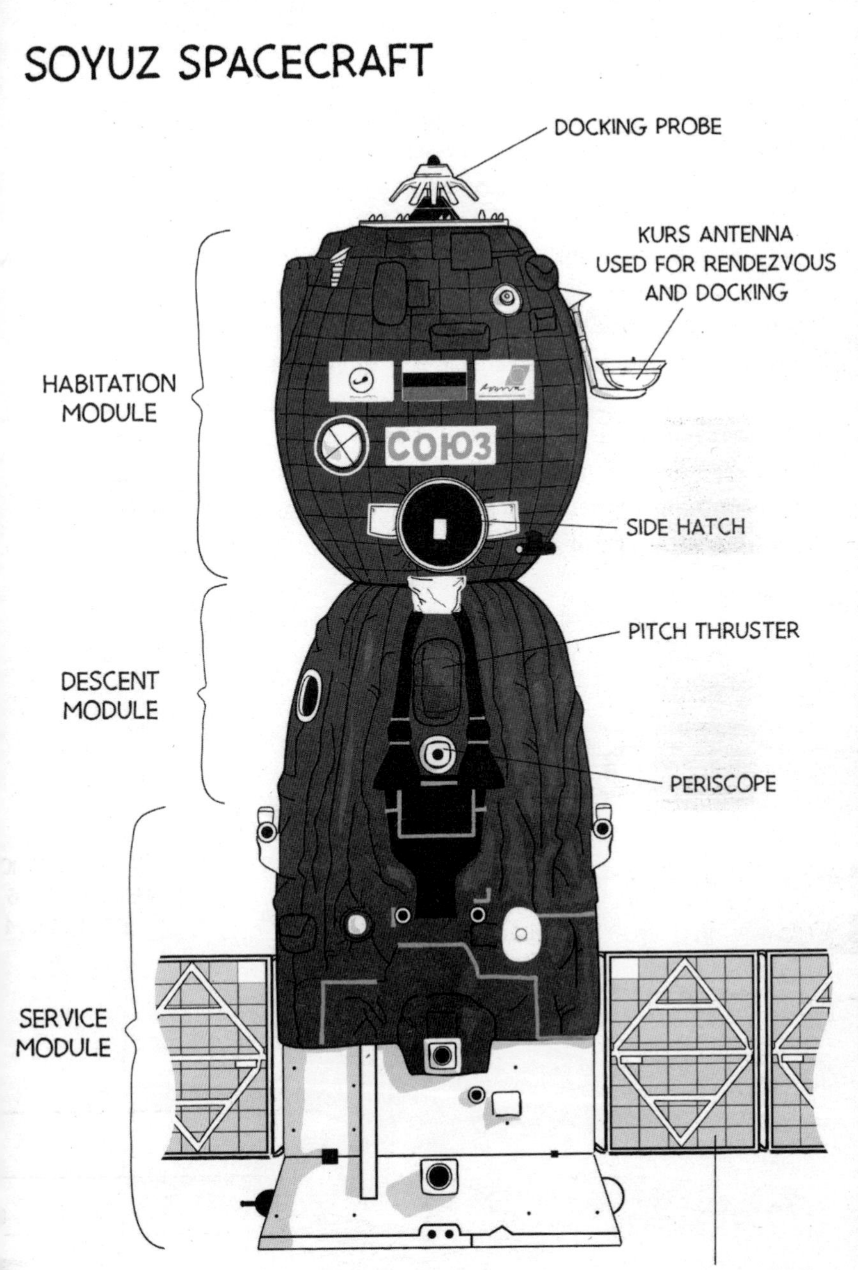
SOYUZ SPACECRAFT
DOCKING PROBE
KURS ANTENNA USED FOR RENDEZVOUS AND DOCKING
HABITATION MODULE
СОЮЗ
SIDE HATCH
PITCH THRUSTER
DESCENT MODULE
PERISCOPE
SERVICE MODULE
SOLAR PANELS

Q *How many 'g's do you experience on launch?*

A The amount of 'g' (or acceleration) experienced during launch depends upon which rocket you're riding – every rocket has its own g-profile, which describes the acceleration that you would feel on your body throughout the entire ride into space. At first glance, it can look a bit messy. Below is the g-profile for our Soyuz TMA-19M:

So why the three peaks on the graph? Well, getting into orbit takes an awful lot of energy. We use rocket fuel to provide this energy, but rocket fuel is heavy and has to be contained within solid structures. Once the fuel has been burnt, we no longer need the structure and so we jettison empty fuel tanks on the way up, in order to reduce weight. This is called 'staging' and the Soyuz rocket has three stages. What this means for the crew is that the 'g' we experience during launch will vary, depending on which stage we are riding and how much fuel we have burnt.

You can see that the greatest acceleration occurred during the first stage. This was when we had all four first-stage boosters firing in

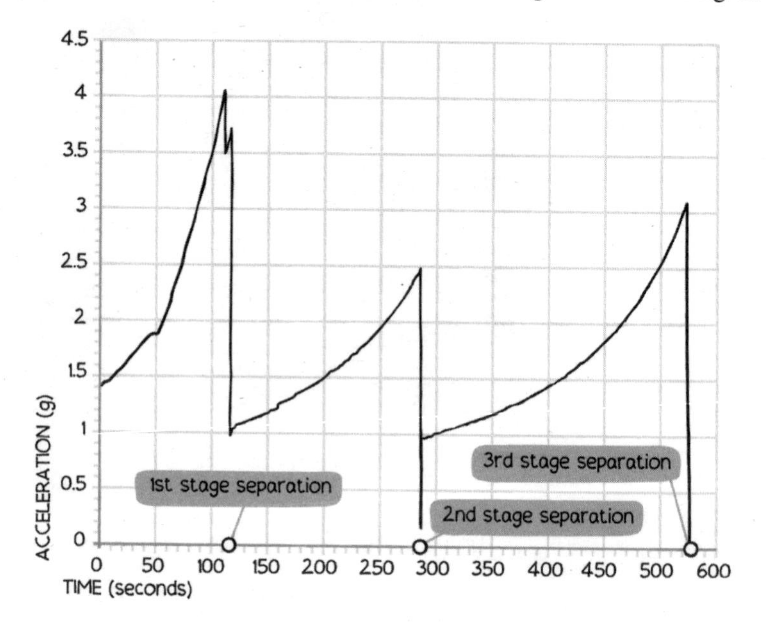

addition to the second stage, delivering about nine million horsepower and accelerating faster than a Formula 1 racing car. As we burnt fuel, the rocket began to get lighter, but it was still producing the same phenomenal amount of thrust, which is why the acceleration kept building, to a peak of just over 4g. This was an amazing feeling, being pinned further and further into my seat, tensing my stomach muscles, concentrating on the good breathing technique that I had been taught a few months earlier in the centrifuge . . . and trying not to laugh out loud at the thrill of it all.

When the first stage was jettisoned, there was a big jolt and a rapid deceleration. Inside the Soyuz, it actually felt like we were being pitched forward, and there was a sensation of falling. Soon the g-load slowly built again as the second stage started to pick up the pace, albeit a much more sedate ride than the first stage. It was during this second stage that I decided to give a 'thumbs-up' to the camera. One of the commander's jobs during launch is to switch cameras so that Mission Control can

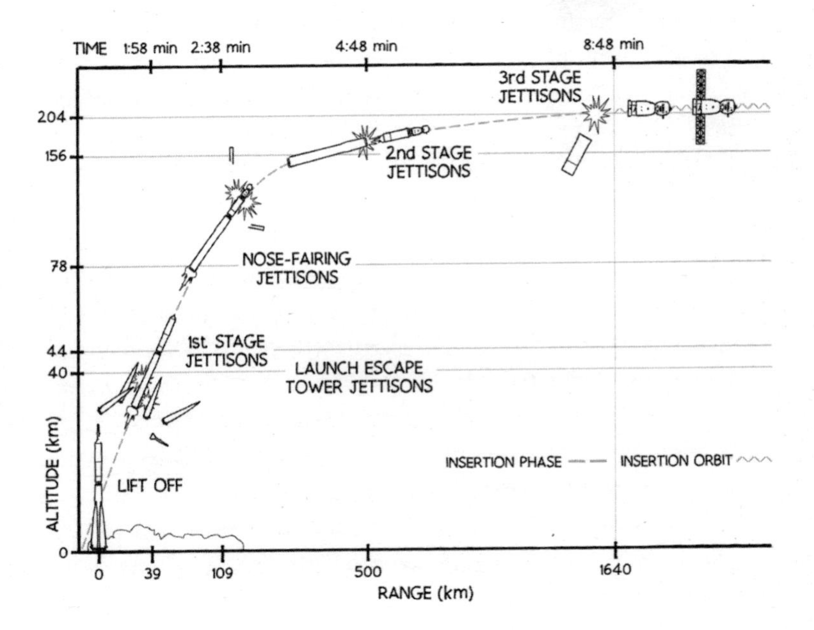

observe all the crew members at different times. As Yuri switched cameras, we were only experiencing 1.5g, which is why it was easy to raise my arm and wave.

Another jolt and the second stage was falling to Earth, leaving behind the final third stage, with our spacecraft perched on top. I thought the third stage was the most exhilarating part of the launch. Although the acceleration wasn't as aggressive as the first stage, the rocket was almost horizontal at this point, having already ascended into space. The feeling of pure speed was overwhelming and I remember thinking, 'How much longer can this possibly go on for?' When the third-stage engine cut out, there was another big jolt, except this time it went eerily quiet and suddenly objects were floating inside the spacecraft – we were in orbit.

Q *When does the sky stop and the atmosphere become space?*

A For most purposes, the official boundary between 'sky', or Earth's atmosphere, and space is at 100 km. This is called the 'Karman line' (named after Theodore von Karman, a Hungarian-American engineer and physicist). But it's not quite as simple as that. Our atmosphere is difficult to measure because it gets progressively thinner as altitude increases.

A distance of 100 km actually puts us in the thermosphere, which extends from 80 km up to 500–1,000 km. So the ISS, orbiting at about 400 km, is also in the thermosphere. Although the ISS is most definitely in 'space', there are still some molecules of gas floating around at 400 km. However, this 'air' is so rarefied that a molecule has to travel about 1 km before bumping into another molecule (that's a pretty lonely molecule, considering there are around $3x10^{22}$ molecules of air in your lungs right now – that's 30,000,000,000,000,000,000,000 molecules!). But the effects of this sparse atmosphere are still enough to create a small amount of drag on the ISS and cause its orbit to slowly decay, on average by 2 km every month. This is why the ISS needs to have an occasional re-boost to raise its orbit and prevent it falling back down to Earth. Other satellites, such as

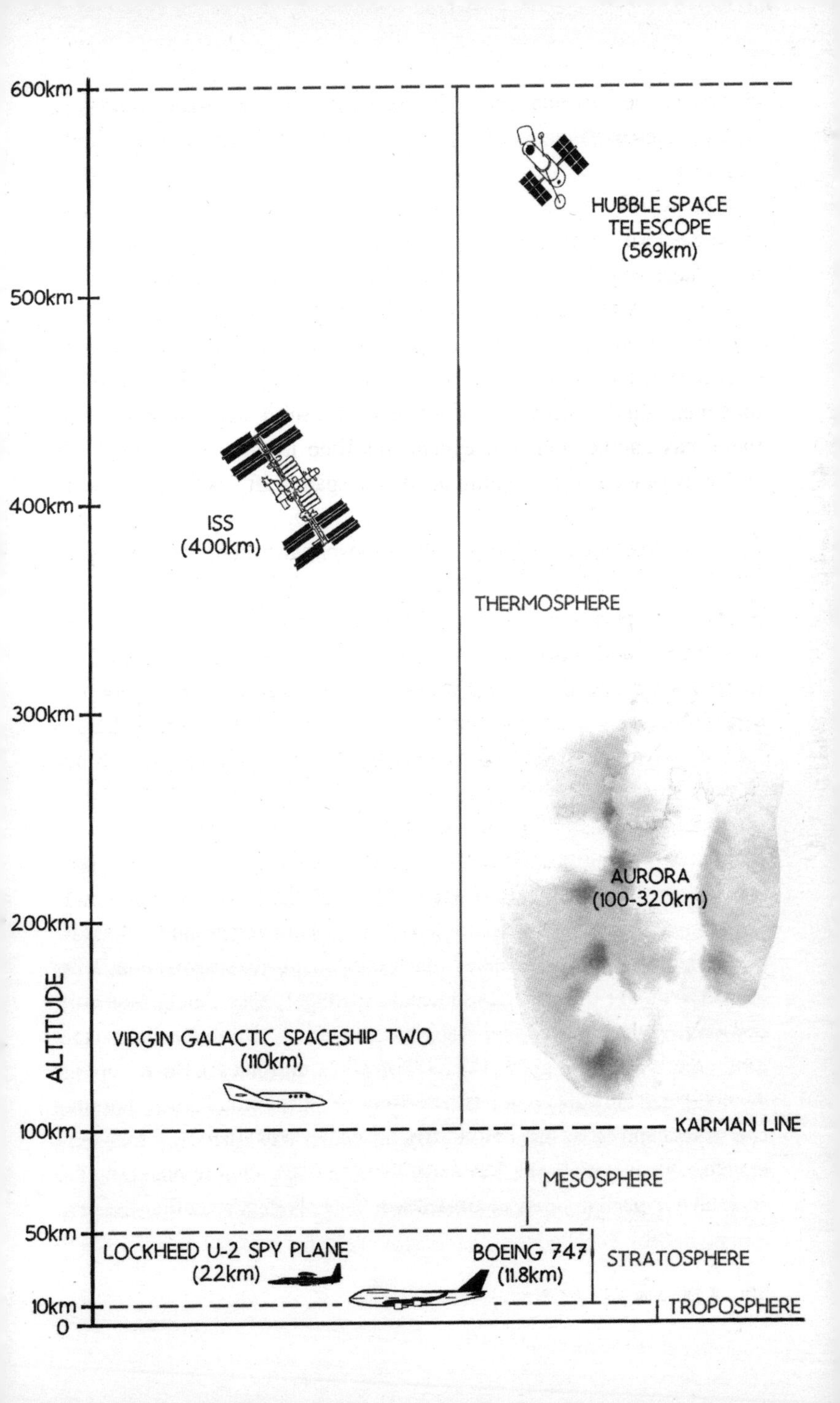
600km
500km
400km
300km
200km
100km
50km
10km
0
ALTITUDE
HUBBLE SPACE TELESCOPE (569km)
ISS (400km)
THERMOSPHERE
AURORA (100-320km)
VIRGIN GALACTIC SPACESHIP TWO (110km)
KARMAN LINE
MESOSPHERE
LOCKHEED U-2 SPY PLANE (22km)
BOEING 747 (11.8km)
STRATOSPHERE
TROPOSPHERE

the Hubble Space Telescope, up at around 560 km, are also experiencing this atmospheric drag and are slowly dropping back down to Earth.

The ISS is also travelling through the ionosphere, which extends throughout the thermosphere and beyond, to the exosphere. The ionosphere is a layer of the atmosphere that has been ionised by solar and cosmic radiation; atoms stripped of electrons, leaving behind a shell of energetic-free electrons and positive ions surrounding Earth. Of benefit to us is the fact that ionised gases are good at refracting high-frequency radio waves, which is why we can bounce radio signals off the ionosphere and communicate with people over large distances (alternatively, of course, you could just use Skype, Facetime, Snapchat . . .).

So, whilst the exosphere extends to a dizzy 10,000 km, where it merges with the solar wind, some scientists think we should say that space begins way back down at 50 km, at the top of the stratosphere, below which we find 99 per cent of the air in our atmosphere. But the International Astronautical Federation decided on the Karman line to mark the beginning of space at 100 km, where Earth's atmosphere is so negligible that conventional aircraft cannot travel fast enough to generate aerodynamic lift.

Q *Why do rockets need to go so fast?*

A Getting into space is one thing – remaining there is another thing altogether. If the Soyuz rocket engine fails during launch at 100 km then, despite having made it to space, you're not going to stay there for very long. That's because you don't yet have enough velocity to remain in orbit. The rocket will instead follow a 'sub-orbital' trajectory, falling back down under the influence of Earth's gravity to hit the ground. The difference with an orbital trajectory is that you still fall back down under the influence of Earth's gravity, but you don't hit the ground, because you're now going so fast that the rate at which you fall to Earth exactly matches the curvature of Earth's surface, and you will remain in orbit for ever, unless acted upon by another force. This magical velocity needed to remain in orbit is called the 'first cosmic velocity' and is 7.9 km/s, which is 28,440 km/h – or about ten times the speed of a bullet . . . and that is why rockets need to go so fast!

Q *How long does it take to get to space? – Jake @trislowe*

A Well, that varies, depending on which rocket you use and, essentially, its 'thrust to weight' ratio. There are other factors involved, of course (drag, dynamic pressure and structural limitations, to name a few), but in a nutshell it's just like any other vehicle – a powerful engine on a strong, light, aerodynamic frame is going to get you there faster. As far as the Soyuz is concerned, if we take the official definition of 'space' as 100 km, then you'll get there in a little over three minutes. By this time you'll already be travelling at several times the speed of sound. The first US astronaut, Alan Shepard, was launched on a Mercury-Redstone rocket on 5 May 1961. This rocket was derived from a US Army ballistic missile and although it was not able to achieve orbital velocity, it was small and light, which meant a very rapid journey into space. Shepard got to 188 km in about two and a half minutes, experiencing 6.3g on the way up. Now that must have been a fun ride!

Q *How long does it take to get to orbit?*

A Once we had passed that momentous 100 km boundary and were officially in 'space', the Soyuz took a little longer to reach its initial orbital insertion at about 230 km. The entire launch sequence took a thrilling 8 minutes and 48 seconds from launch pad to orbit – which may seem like a quick journey to space, but I can promise you it felt like an awfully long time sitting on top of a speeding bullet!

Q *What do astronauts actually do during launch – are you flying the spacecraft or is it done by computers?*

A During launch, the crew's primary focus is to monitor all the systems and ensure that everything is functioning normally. The entire launch process is automated, and only in the event of an emergency does the crew need to intervene.

Apart from the rocket staging described earlier, there are a couple of other events that we were looking out for during launch. One of these was the jettison of the nose-fairing, which protects the Soyuz spacecraft beneath it. Once the rocket had reached about 80 km altitude, it had climbed above most of the air in the atmosphere. There was now very little drag and therefore very little aerodynamic heating, caused by skin friction (by which I mean friction against the outer surface or 'skin' of the rocket) from high-speed collisions with air molecules. At this point the nose-fairing had done its job in protecting the spacecraft during its passage through the lower atmosphere and it was now dead weight – time to jettison it. This was a memorable moment, when suddenly we could see out of our windows for the first time as the nose-fairing blew away from the spacecraft. Of course we were still strapped tightly into our seats and the windows were not quite at eye level, so it wasn't a perfect view. Nonetheless, looking up, we could clearly see the sky changing rapidly from blue to black as we left the remains of the thin atmosphere behind us and headed into space.

At this point I was cognisant of checking the pressure inside the spacecraft. From the right seat it was hard to see some of the displays on the control panel, but among the things I was able to monitor were the life-support systems and internal pressure. Since we were rapidly approaching a vacuum, it was a good time to check the integrity of the spacecraft. Towards the end of the launch we were all closely watching the clock, in anticipation of the third-stage separation. This came as a big jolt, as the engine cut out and the Soyuz spacecraft was severed from the upper stage of the rocket. Apart from the jolt, there are several indications inside the capsule that confirm successful orbital insertion. If these do not occur, then the crew has only a matter of seconds to intervene. Thankfully, we had a clean third-stage separation and a good orbital insertion – and then, with no time to waste, we were straight into the checklist to prepare for an engine burn to begin the rendezvous sequence with the ISS.

Q *What happens if something goes wrong during launch?*

A The Soyuz is one of the most reliable rockets to fly to space. It's also one of the safest. Having said that, getting to space is not easy, and there have been problems in the past. The good thing about the Soyuz is that there is a launch escape system that enables a survivable return to Earth, if something goes wrong at any point from the launch pad to orbit. I say 'survivable' and not 'safe' because the crew could be exposed to acceleration forces in excess of 20g during some phases of an aborted launch – and there's nothing safe about 20g.

There is a group of small rocket thrusters mounted at the top of the rocket – namely, the launch escape tower. In the early stages of a launch, their job is to pull the nose-fairing, along with the descent module (plus crew) and the habitation module, away from the rest of the rocket in the event of an emergency, then get the spacecraft to a safe height so that the parachute can operate normally. The launch escape tower is only needed for the first 1 minute and 54 seconds of launch. After this, the rocket will be about 40 km high – safe enough for the parachute system to work – and so the launch escape tower is jettisoned, followed a few moments later by the nose-fairing.

After this, if anything goes wrong, then the automatic abort system will shut down the rocket engine and separate the crew compartments, once again enabling the descent module's parachute and landing systems to function normally. Throughout the launch the indication to the crew that something has gone badly wrong (other than a massive noise, vibration or explosion, of course) is a red warning light 'АВАРИЯ НОСИТЕЛЯ' (booster emergency) – it's the first emergency light on the warning panel and one that no crew member ever wants to see.

The launch escape system saved the lives of Russian cosmonauts Vladimir Titov and Gennady Strekalov on 26 September 1983. A problem occurred during the final stages of fuelling, just prior to launch, and a fire began at the base of the rocket. Although there was some delay in operating the launch escape system (the control cables had burnt through and the 'Abort' command had to be issued by radio link), it was finally activated with only seconds to spare before the rocket blew up, and the crew were launched skywards at 14–17g for five seconds, landing about 4 km away. Years later, during an interview, Titov claimed that one of the crew's first actions following the event was to deactivate the cockpit voice recorder, due to the excessive amount of swearing that had occurred!

Another emergency situation developed on 5 April 1975. Commander Vasili Lazarev and Flight Engineer Oleg Makarov launched on Soyuz 18A, headed for the Salyut 4 Space Station. At an altitude of 145 km, the second and third stages were supposed to separate. However, only three of the six locks that held the two stages together released, and so the third stage ignited whilst still attached to the second stage. The thrust from the third-stage engine eventually broke the remaining locks, but the spacecraft was off-trajectory and an automatic abort sequence was initiated. An abort from such a high altitude will result in a steep re-entry angle and the crew experienced very high deceleration forces of up to 21.3g. Whilst Makarov made a full recovery and flew two more missions, Lazarev suffered internal injuries and never flew again.

Q *Where would you land if the launch was aborted?*

A An important factor to consider, when launching rockets, is the terrain that you are going to be flying over. The requirement for rocket staging means that there will be debris falling from space and, in the event of an aborted launch, a spacecraft, too.

Launching due east from Baikonur would be the most efficient, in terms of maximising Earth's 'free' rotational velocity, but it would also drop the first-stage boosters on China, not to mention complicating any search-and-rescue mission in the event of an aborted launch. Therefore, in order to keep the majority of the launch trajectory over Russian territory, the Russians aim off a bit to the north (I may be trivialising the amount of maths involved in this 'aiming off').

So, back to the question. If the launch was aborted, you would probably land somewhere in Kazakhstan, eastern Russia or, if the abort occurred shortly before getting to orbit, you might make it all the way

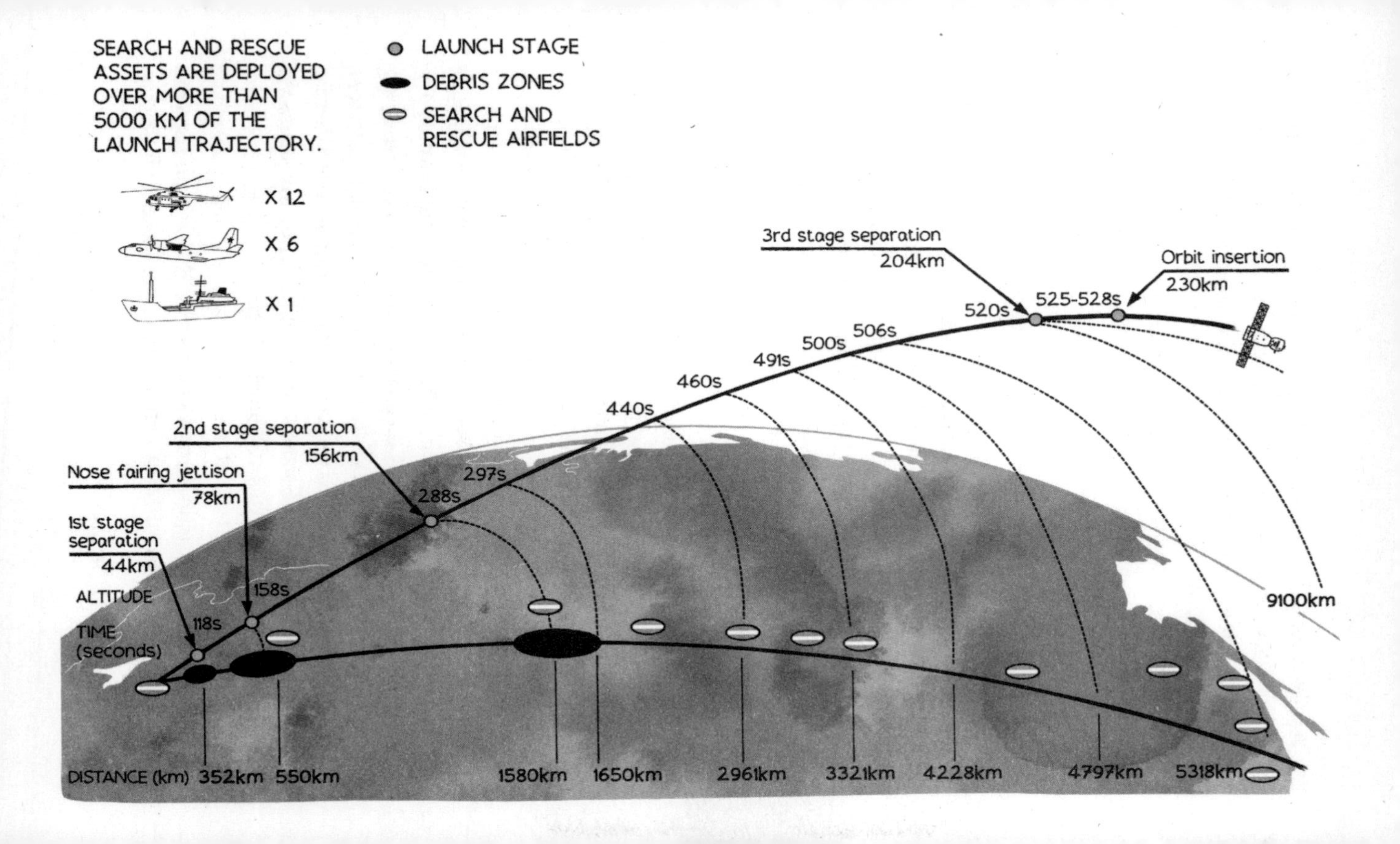
SEARCH AND RESCUE ASSETS ARE DEPLOYED OVER MORE THAN 5000 KM OF THE LAUNCH TRAJECTORY.
X 12
X 6
X 1
LAUNCH STAGE
DEBRIS ZONES
SEARCH AND RESCUE AIRFIELDS
3rd stage separation
204km
Orbit insertion
230km
520s
525-528s
506s
500s
491s
460s
440s
2nd stage separation
156km
Nose fairing jettison
78km
297s
288s
1st stage separation
44km
ALTITUDE
158s
118s
TIME (seconds)
9100km
DISTANCE (km)
352km
550km
1580km
1650km
2961km
3321km
4228km
4797km
5318km

to the Sea of Japan. Roughly speaking, an abort during the first 283 seconds would put you over the flat plains of eastern Kazakhstan; from 283 to 492 seconds, you would land in mountainous terrain – the region of southeast Russia that shares a border with Mongolia. Then there is a 14-second abort period that would actually put you in the very northernmost part of China. Finally, from 506 seconds until orbit is achieved at 528 seconds, you would be getting your feet wet in the Sea of Japan.

Clearly, this covers a vast area, which is why the logistics surrounding the search-and-rescue operations for a Soyuz launch are phenomenal. The Russian Federal Air Transport Agency is responsible for providing these Air and Space Search and Rescue services. The chart on page 37 shows the scale of preparations for a Soyuz launch, involving 18 aircraft deployed from 12 airfields covering more than 5,000 km – and a maritime recovery vessel in the Sea of Japan.

Q *How long does it take to get to the ISS?*

A It used to take just over two days (34 orbits of planet Earth, to be exact) to get to the ISS. This meant that, once in space, the crew opened the hatch to the habitation module and used the extra room for sleeping, eating and going to the loo. There's plenty of drinking water and dry-food rations on board the Soyuz, but not a lot of in-flight entertainment. Thankfully, the magnificent views from space help to pass the time.

However, such a long rendezvous is cramped, uncomfortable and not the most efficient use of time for astronauts or Mission Control, and so in August 2012 a new four-orbit rendezvous profile was attempted, using a Progress cargo vehicle. This fast rendezvous had not been attempted before, as it required very accurate orbital insertion, by which I mean a precise launch and trajectory into orbit. The new Soyuz-FG launcher and Soyuz TMA-M series spacecraft were able to achieve such precision.

The big change that allowed this to happen was to pre-program the first two engine-burns after launch into the spacecraft's guidance, navigation and control computer. With a long rendezvous, Mission Control would have observed the Soyuz trajectory following launch to check its orbital parameters, then sent up the commands for subsequent engine-burns, correcting any errors that may have occurred during launch. The new way of thinking accepted that although there may have been minor errors in the initial orbit insertion, these could be accounted for later on, in the rendezvous sequence. By starting these first two engine-burns early, it enabled the Soyuz to arrive at the ISS just six hours after launch.

In addition, the approach and docking usually take another 30 minutes and then there are several checks to complete, prior to opening the hatch to the space station. In total, the crew usually arrive on board the ISS about eight to nine hours after launch. I was fortunate to launch on a 'short' (four-orbit) rendezvous, but the two-day rendezvous is still used if a new version of a rocket or spacecraft software is being tested and, of course, as a backup option if something goes wrong during the short rendezvous sequence.

Crew fatigue is also a consideration with the 'short' rendezvous. Launch day can be pretty exhausting – the crew will have been awake for about nine hours prior to launch, with the critical docking taking place around 15 hours into the day, with several hours of demanding tasks still remaining after that. I certainly remember sleeping extremely well when I finally retired to my crew quarters on the first 'night' aboard the ISS.

Q *How do you rendezvous with the ISS?*

A Our Soyuz was launched into a very low (about 230 km) and slightly elliptical orbit. This is so low that there is still a small amount of atmospheric drag and, without any additional manoeuvres, we would have completed only about 20 orbits before re-entering Earth's atmosphere. So first we had to 'normalise' our orbit (make it a bit more circular) and raise our altitude to about 340 km. For our 'short' (four-orbit) rendezvous, this was accomplished during two 'pre-programmed' engine-burns, using something called a 'Hohmann Transfer', which is the most fuel-efficient method of getting from one circular orbit to another.

Without going too deeply into orbital mechanics, if you fire the engine on a spacecraft in orbit around Earth, you will not actually speed up. What happens is that you go higher (farther into space) and slow down. This sends you into an elliptical orbit, and you reach the highest part of the orbit on the opposite side of the planet from where you did the engine-burn. Do nothing else and you will fall back down towards your starting point, speeding up again in the process. So the trick is to fire the engine a second time when you get to the highest point – and hey presto, you achieve a nice circular orbit at a higher altitude.

This higher orbit is called the 'phasing orbit'. Mission Control could now take a look at our phasing orbit and work out what errors might have occurred up until that point. Corrections were sent up to the guidance, navigation and control computer, and a further two engine-burns were completed to make the necessary adjustments to our phasing orbit and

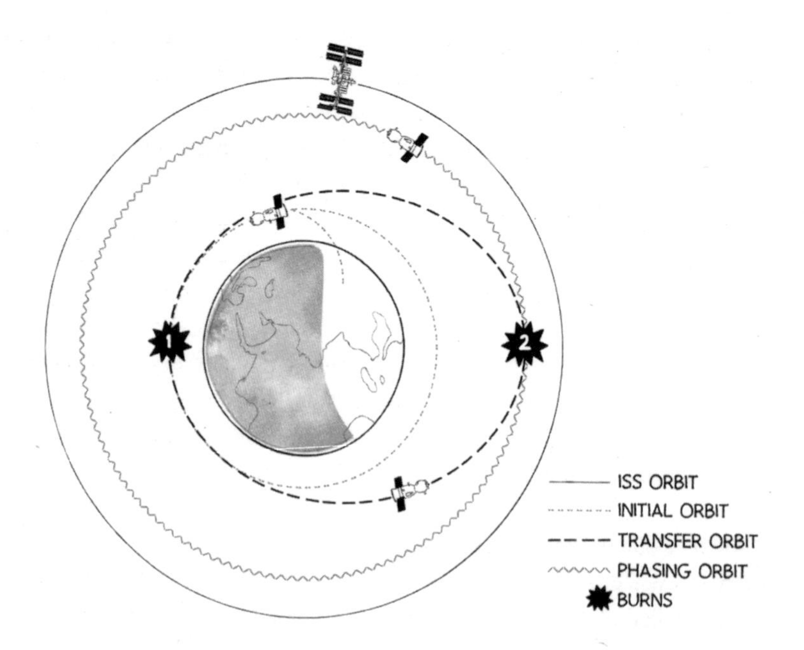

raise us to about 370 km. These engine-burns were not aggressive (most lasting less than a minute) and we felt only a very small acceleration. Instead, when the engine fired, it sounded like a muffled rumble and we were simply pushed gently back into our seats.

Next, we completed a three-engine-burn that took us from our phasing orbit and raised our altitude to match that of the ISS (around 400 km). The third burn was actually a series of small burns, and at this point we turned the spacecraft around and fired the engine in the opposite direction. This is called the 'braking parabola' and, following this series of burns, our Soyuz had closed to within about 150 metres of the ISS and we commenced a 'fly-around' until it was aligned with the correct docking port. One of the most incredible sights during this phase of the rendezvous was watching the ISS transition from being a small, bright dot in space to this huge structure looming out of the darkness, many times the size of our tiny Soyuz spacecraft. I couldn't help but smile, as I recalled the scene in the James Bond film *Moonraker* when Drax's secret space station appears in a similar manner.

The last phase of the rendezvous is the final approach. All being well, the spacecraft's probe should be lined up with the designated docking port, and then the Soyuz closes until it's captured by the ISS. When the probe enters the docking cone, the Soyuz thrusters give a small boost to ensure a firm capture. The probe then retracts to bring the two vehicles together and, finally, hooks close to secure the Soyuz to its new home for the next six months. Although the crew is extremely busy during the six-hour rendezvous, all engine-burns, manoeuvring and docking are fully automated, with no requirement for the commander to actually 'fly' the spacecraft. However, in our case things didn't quite go according to plan, which brings me to the next question . . .

Q *What was your scariest moment in space?*

A During our fully automated docking sequence there was a problem. We were approaching from below the space station, docking to a Russian module called *Rassvet*. I was struck by the sheer size of the ISS as the

Soyuz crept slowly inside the enormous solar panels – I remember commenting to Tim Kopra on the size of those solar panels, forgetting that at this stage everything we said was being transmitted back to Mission Control . . . rookie error! As we edged nearer the space station, I was surprised to see quite how close the Cygnus resupply spacecraft looked; it was docked just forward of our docking port and appeared large outside my right-seat window. Cygnus had two wide umbrella-shaped solar panels protruding on either side at the base of the vehicle, and it seemed there would be only a metre or so of clearance as our Soyuz nudged closer.

Everything was progressing normally until, at just 17 metres away from the ISS, one of our thruster pressure-sensors failed, forcing the Soyuz to command an automatic abort and sending us back out into space. Yuri, our very experienced Russian Soyuz commander, was in the centre seat and on his sixth mission to space. He quickly took manual control of the spacecraft. Using two hand-controllers, he had to realign the spacecraft and fly it back in to dock with the space station. However, we were about to transition from day to night, with only three minutes remaining until we entered the shadow of Earth. The Sun was very low and reflecting off the space station towards the Soyuz, which made it almost impossible for Yuri to see *Rassvet*'s docking target clearly.

During Yuri's first docking attempt, as the Soyuz approached the ISS, it drifted towards the aft end of the space station and yawed off-target. Docking is one of the most critical phases of spaceflight. A collision between vehicles can not only cause irreparable damage, potentially leaving a spacecraft stranded with no power or control, but it can also rupture the hull, causing a rapid depressurisation and placing the crew's lives in jeopardy. Such a situation occurred on 25 June 1997 when a Progress resupply vehicle struck the Mir space station (see page 198 for more detail on this, when I answer the question about what would happen if the ISS was hit by space debris).

Thankfully, Yuri's experience and hours of manual docking attempts in the simulator paid off. Recognising the danger, he backed away from the space station, realigned the spacecraft and manoeuvred the Soyuz back in, for a textbook docking. That was probably the moment of greatest apprehension for me, Tim and Yuri – getting that docking right and ensuring we arrived safely aboard the space station.

Fortunately there weren't any other scary moments during the mission. Scary moments in space are never a good thing! However, some parts of spaceflight definitely have more potential for things to go wrong, and we would always focus our training on these 'high-risk' areas, as a means of mitigating the dangers. Unsurprisingly, the main areas that require greater attention to detail are launch, re-entry, spacewalking and docking operations (including visiting cargo vehicles).

Q *What surprised you the most when you first got into space?*

A During my first orbit of planet Earth, I looked out of my Soyuz right-seat window and the thing that surprised me the most was just how *black* space appeared during the daytime. It's the blackest black you could possibly imagine, and it looks truly remarkable. I think this is because we are so used to looking at the night sky here on Earth and seeing stars, diffused sunlight or cloud cover reflecting terrestrial light sources. Even on the darkest of nights our atmosphere gives a faint emission of light, called 'airglow', and this optical phenomenon causes the night sky never to be completely dark.

In space it is very different. During the daytime the Sun easily outshines the brightest of stars and planets, and our eyes adjust their sensitivity to accommodate this brightness. When I looked out into space I was simply rewarded with a vista of black ink. During my spacewalk it felt almost intimidating being on the farthest edge of the space station and having this dark abyss of space lurking over my right shoulder – it was certainly a good incentive not to let go!

Q *Did you feel unwell when you first got to space?*

A During the first 24 hours in space most astronauts will experience some dizziness, disorientation . . . and perhaps the occasional stomach-emptying. I felt fine during launch and during the entire six-hour journey to the space station. I remember unstrapping my harness and enjoying those first feelings of weightlessness, albeit in the confines of the Soyuz. Yuri's non-verbal skills are impeccable. I've never known anyone who can convey so much simply with a look, without saying anything. As I floated up out of my seat, he gave me a glance that said, 'Just take it slowly – early days.' That turned out to be good advice. When I first arrived on the ISS, I felt a bit disorientated, but otherwise fine – no nausea. But the next day I felt a bit rough. This wasn't debilitating or incapacitating, as seasickness can sometimes be. Instead, one minute I'd be feeling perfectly fine and then suddenly I'd have five minutes of feeling dizzy and nauseous, before being able to get straight back to work again.

The main culprits for causing this dizziness (or vertigo) and disorientation are our ears. On Earth, our inner ear uses fluid (endolymph) in our semicircular canal to detect rotations of our head. This is part of the vestibular system, the other part being otolithic organs, which sense linear accelerations and are very sensitive to gravity and motion. When we tilt our head, it's the otolithic organ that senses the change in the gravity vector and sends this information to the brain.

Now, if we inject microgravity – suddenly the endolymph is in freefall and the otoliths have no gravity vector! Since the vestibular system provides powerful stimuli to the brain, in order to determine balance and spatial orientation, these inputs can be suddenly overwhelming. At the same time the brain also receives visual and proprioceptive (relating to position and movement) signals and when this information does not match up with the erroneous vestibular system, you have a perfect recipe for motion sickness.

However, the brain is brilliant at adapting to new environments. In my case, I felt that once my body learnt to ignore the confusing signals coming from my vestibular system, life was great. It was like the flip of a light switch and when I woke up on the second day, I felt perfectly fine from that moment on. In fact towards the end of the mission I attempted to make myself dizzy. I curled into a ball and asked Tim Kopra to spin me round at a very fast rate for a couple of minutes, whilst I moved my head in different directions, in an attempt to provoke a reaction. This is something that would have made me feel incredibly dizzy and nauseous on Earth, but since I had fully adapted to living in space, it had very little effect on me and I was surprised that I could barely make myself dizzy at all.

Q *Who was the first person to greet you on the ISS when you opened the hatch?*

A Once we had successfully docked to the space station, it took nearly two and a half hours to pressurise the docking port, complete leak checks, place the Soyuz into hibernation mode, change into our flight suits and prepare for hatch-opening. During this time there was quite a bit of communication going on between Yuri and the Russian crew on the space station (Sergey Volkov and Mischa Kornienko), since many of the checks have to be coordinated. However, at some point in that process a familiar New Jersey accent came over the radio. ISS commander Scott Kelly welcomed us to space and then casually asked what we would like for dinner. Scott had rummaged through some of our 'bonus food' containers and had picked out a few items that he was going to put in the food warmer, to be ready on hatch-opening.

Having just lived through my first rocket launch, rendezvous and an adrenaline-filled docking, it suddenly felt like I had come all that way simply to place an order at a drive-through. I asked Scott for a bacon

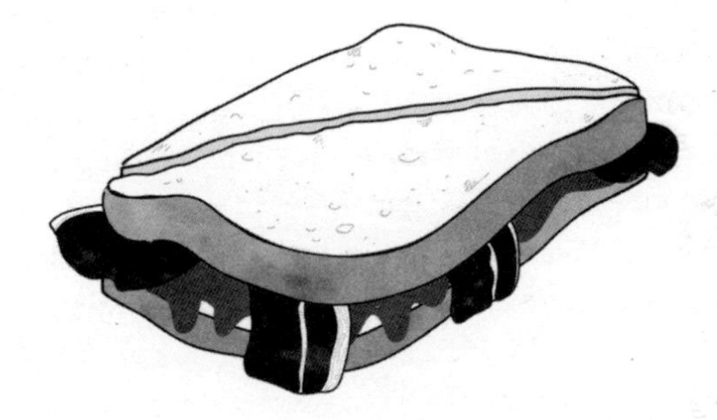

sandwich and couldn't help but smile at how bizarre the whole situation felt. Shortly afterwards the hatch opened and Sergey, Mischa and Scott (in that order) were there to welcome us aboard, with huge grins, to begin our six months of life and work on the International Space Station.

Before we explore life and work on the ISS in more detail, we're going to go back a bit, to examine the training to become an astronaut. What does it really take to have the *right stuff* for today's missions to space? You might be surprised.

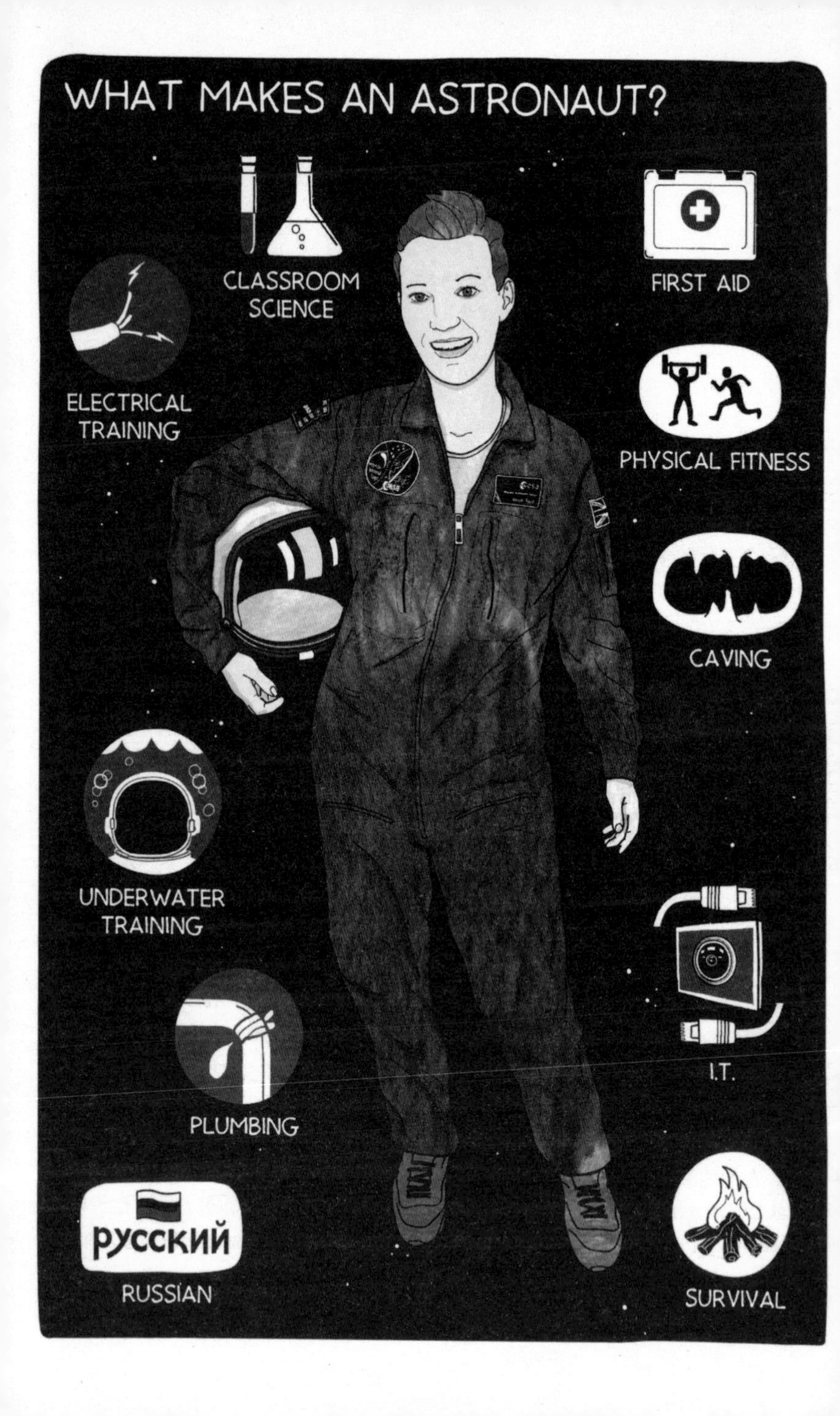
WHAT MAKES AN ASTRONAUT?
CLASSROOM SCIENCE
FIRST AID
ELECTRICAL TRAINING
PHYSICAL FITNESS
CAVING
UNDERWATER TRAINING
I.T.
PLUMBING
РУССКИЙ
RUSSIAN
SURVIVAL

TRAINING

Q *My oldest child (who has career ambitions of being an astronaut) would like to know: how, when and why did you decide to become an astronaut? – Amanda Lou*

A That's a lot to cover! In a nutshell, here is a short history of my route to becoming an astronaut. I hope this entire chapter will also give you a real flavour of what it takes to pass the rigorous astronaut-selection process and the extensive training and preparation that are required ahead of a mission to space. As you will see, there's no single route to becoming an astronaut, but there are some areas to be aware of that may help to maximise one's chances. Good luck!

1972: Early life

My father had always been interested in historical aircraft and took me to air shows from an early age. I was hooked from the very beginning – thrilled by the noise and the daring flying displays, whilst marvelling at the machines themselves, wondering what made them fly and why they were so varied in design.

I was also fascinated by the stars and the universe. I loved looking up at the bright strip of the Milky Way and being able to pick out some of the major constellations. However, when it came to deciding my first choice of career, it was neither astronomy nor becoming an astronaut that was my driving passion – I wanted to fly. I loved everything about flying (I still do!) and could not wait to train as a pilot. At high school I enjoyed studying maths, science and graphic design, but outside the classroom it was being a member of the school's Combined Cadet Force that really shaped my early years. Although I was drawn to the Army section, as opposed to the Royal Air Force section, I would always volunteer to go flying at every opportunity and I savoured my early forays in gliders and small, powered fixed-wing aircraft.

1994: Getting my wings

As someone who was more comfortable wearing Army uniform, but with a burning passion for flying, it was no surprise that I set my sights on joining the Army Air Corps (AAC) at the age of 19. Such was my enthusiasm for flying that I decided to bypass university and instead entered the Royal Military Academy Sandhurst, graduating in 1992 as a Second Lieutenant. I began pilot training soon afterwards, first learning to fly the de Havilland Chipmunk. Built during the 1940s, this tandem-seat (one person behind the other), single-engine aircraft was often used to train pilots. I found the seating and tail-wheel arrangement somewhat reminiscent of a Second World War fighter, and it was a joy every time I flew in one. Having progressed to flying helicopters, I was awarded my wings in 1994 and shortly thereafter began an exciting four years flying reconnaissance missions all over the world, including operations in Bosnia during the 1990s Balkan War.

My aviation career rapidly progressed from being a reconnaissance pilot to becoming an instructor pilot, teaching new students how to fly. It was during this time that I was offered an amazing chance – to spend three years serving with the US Army's 1st Cavalry Division, flying Apache helicopters in Texas. If you've ever seen the 1979 movie *Apocalypse Now*, you may recall the surf-loving 1st Cav pilots flying low into battle, with Wagner's 'Ride of the Valkyries' blaring out over loudspeakers. They seemed like an interesting bunch, and I didn't need much persuading to pack my bags and head to the USA.

This was in 1999, prior to the Apache helicopter entering service with the British Army, and so it was a golden opportunity for me to learn everything I could about this new aircraft. On my return to the UK, I was promoted to the rank of Major and spent the next three years training British Army pilots how to fly and fight in this incredibly capable machine.

2005: Test pilot

In 2005, an opportunity opened up – one that was to set me squarely on the path to becoming an astronaut, although I didn't know it at the time. Throughout my aviation career I had always been interested in testing the theory behind flying. I loved learning about new systems, discovering how aircraft really worked and exploring the boundaries of their performance. This was the work of test pilots, and so I set my sights on trying to join their ranks. I studied hard to pass the demanding selection for test-pilot training and embarked on an intensive year-long course, flying more than 30 types of aircraft, including helicopters, fast jets, heavy transport and pretty much anything our instructors could find for us that could fly. On graduating from Empire Test Pilots' School at Boscombe Down, I became the senior test pilot for the Apache with the Rotary Wing Test Squadron – just as that particular helicopter began to be used in Afghanistan. It was a hugely rewarding time, knowing that frontline pilots were benefiting from the work we were doing, but most of all I loved pushing aircraft to their limits. In some experiences I had I was genuinely taking the aircraft somewhere nobody had taken it before, in terms of speed, altitude and manoeuvrability.

2006: Getting a degree

Test-pilot training involved a significant amount of academic work, in addition to demanding flying exercises. I had never been strong at mathematics, and so my first month was spent burning the midnight oil, bringing my maths up to first-year degree-level standard. I decided that, in addition to catching up on the maths, this was the right time to fill another gap in my education and gain a degree. I enrolled in Portsmouth

University's Bachelor of Science course in Flight Dynamics and Evaluation. As I would later discover, it was a combination of test-pilot training and having degree-level education that really opened the door for astronaut selection a few years later.

As a test pilot, you work very closely with the commercial aerospace sector. Part of your job is to broaden your knowledge and experience, learning about technologies to innovate and improve capability. Space is one of the most demanding environments that humans have ever lived and worked in, and so it was no surprise that, as a test pilot, I began to look at the space sector more closely with a keen interest in those cutting-edge technologies that were being used to drive scientific research and exploration off the planet.

2008: Right place, right time

There are some things in life where good timing is everything, and in that respect I consider myself extremely fortunate. When the European Space Agency held its selection for astronauts in 2008, you could either apply as a pilot with more than 1,000 hours' flight time, or you could apply with degree-level academic qualifications in other fields. I was an experienced test pilot with more than 3,000 hours' flying, a degree in Flight Dynamics and an unquenchable thirst for science, technology and exploration. Like so many others, I jumped at the opportunity. For me, becoming an astronaut was the pinnacle of what a test pilot could aspire to. To be a part of that small team of men and women fortunate enough to venture into space, pushing the boundaries of science, technology and exploration, really was the opportunity of a lifetime – and I was in the right place at the right time.

Q *How did your skills as a pilot transfer to your career as an astronaut?*

A I loved my career as a helicopter pilot, but it certainly came with its fair share of risks. Flying missions at night and in bad weather, or pushing experimental aircraft to their limits, can at times be hazardous.

Dealing with an emergency 60 metres above the ground is not that different from dealing with an emergency 400 km up in space; you need to stay calm, identify the problem and find a solution quickly.

Of course flying a helicopter isn't the same as living on a space station. But my years of dealing with the unknown helped me to prepare for it. Prior to any test flight we would spend hours analysing the hazards, identifying what could be done to mitigate those risks and then training for every eventuality, should something go wrong. That same mindset is needed in space.

Another example of skill transfer is communication. Astronauts depend on good communication with Mission Control on a daily basis to deal with routine tasks, maintain efficiency and prevent errors. In an emergency situation this skill becomes critical in order to prevent a catastrophe. Pilots are very familiar with the way this works. Without clear, concise communications, there's no way a pilot and crew can operate the aircraft properly. The tandem-seat arrangement inside many aircraft requires even greater attention to communication skills, since you lose those often-instinctive 'non-verbal' cues when you can't see the other person, if they're in front of or behind you. Operating equipment is also a transferable skill in itself. For example, later in my astronaut training I learnt how to operate the space station's robotic arm, which is used to manipulate objects in space, including grappling visiting cargo vehicles. This required a high degree of coordination and spatial awareness, with each hand controlling motion in different axes. I actually found this very similar to flying a helicopter.

Of course the most transferrable skill from being a pilot to being an astronaut is that we must learn to fly the spacecraft manually, in case the automatic control systems ever let us down. It's no wonder, then, that most astronauts today (including those not selected from a piloting background) maintain some sort of current flying practice as part of their training and readiness for spaceflight.

Q *Are you more likely to become an astronaut if you join the military as a pilot or if you are a scientist?*

A This is an interesting question and one where the answer has changed over the years. When talking of astronaut selection, people often think back to the early days of human spaceflight. The first Russian cosmonauts and the US Mercury, Gemini and Apollo crews had many similar characteristics, with the vast majority of them being chosen on the basis of their expertise as fighter pilots. The earliest exception to this rule was Valentina Tereshkova, the first woman in space. Prior to her recruitment as a cosmonaut in 1962, Tereshkova worked in a textile factory and was an amateur skydiver. As the missions and objectives in space have changed over the years, so have the astronaut-selection criteria. It is still necessary to have the coordination, spatial awareness and time-critical decision-making skills typical of a pilot, but as an astronaut today you will spend a great deal of your time in space conducting scientific research, maintaining the space station and getting on with your crewmates. That requires a more diverse skill-set.

Today you are equally likely to become an astronaut having trained as a scientist or a pilot. Of the 20 astronaut candidates selected in 2009 by ESA, NASA, the Canadian Space Agency (CSA) and the Japanese Space Agency (JAXA), exactly half did not have a background as a military pilot. In fact, as I mentioned in my introduction to this book, you can become an astronaut as a school teacher, engineer or medical doctor, or from any number of varied careers. One of NASA's latest astronaut recruits had previous experience as an ice-driller and commercial fisherman.

I think one of the most important pieces of advice, for anyone considering a career as an astronaut, comes from ESA's website: 'Above all: no matter what you have studied, you should be good at it.' As we shall see from the next question, your academic credentials or flying experience may get you as far as the interview, but it will be your drive, enthusiasm, character and personality that will secure you the job as an astronaut.

Q *What separated you from the other candidates who applied to be an astronaut?*

A This is a good question and one that I've asked myself. Ultimately, astronauts need to possess a number of skills. Some of these are natural abilities, such as coordination, spatial awareness, memory retention and concentration. However, as missions to space have become longer in duration, other qualities have become equally important, such as communication, teamwork, decision-making, leadership/followership and the ability to work under stress to solve problems. I was fortunate to have had the opportunity to develop many of these skills during my military career, having worked in stressful environments, having undergone extensive leadership training and having practised good crew communication as a pilot for many years.

Each space agency has developed a slightly different selection process, designed to distinguish the broad spectrum of qualities expected of an astronaut. Unusually, when I applied to be an astronaut in 2008, ESA, NASA, CSA and JAXA were all recruiting new candidates, so the differences between each agency's selection criteria were thrown into starker contrast. For example, the Canadians had applicants diving to retrieve bricks from the bottom of a pool, fighting fires and – a particularly demanding stress test – working as a team to plug leaks in a room that was rapidly flooding with cold Atlantic water! In contrast, ESA's selection process was lighter on physical activity, but it did involve a high degree of cognitive testing and psychological profiling, to ensure that astronauts had the right disposition to head into space for several months at a time.

In addition to stringent medical requirements and several rounds of interviews, the selection process tested for a basic level of knowledge in areas such as maths, science, engineering and English language. These evaluations were designed to be stressful, with a minimum of breaks in between tests and a high degree of speed and accuracy required in order to pass.

So to try and answer your question more directly: you don't have to excel in any one area during astronaut selection – provided you have the ability to pass each of the tests, then what will separate you from the other candidates are your personality and character. Having a breadth of different experiences, including working in an international environment, will always be a strong asset, as will an aptitude for foreign languages. After I was selected, I asked an astronaut who had conducted some of the interviews how they had made their choice. The reply was simple: 'I just asked myself: would I like to go to space with this person?'

QUICK QUIZ

Here's a mental challenge for you – this was one of the questions during my selection process. Imagine that you are facing a cube. This cube can roll to the left, right, forward (towards you) or backwards (away from you). There is a dot on the bottom of the cube.

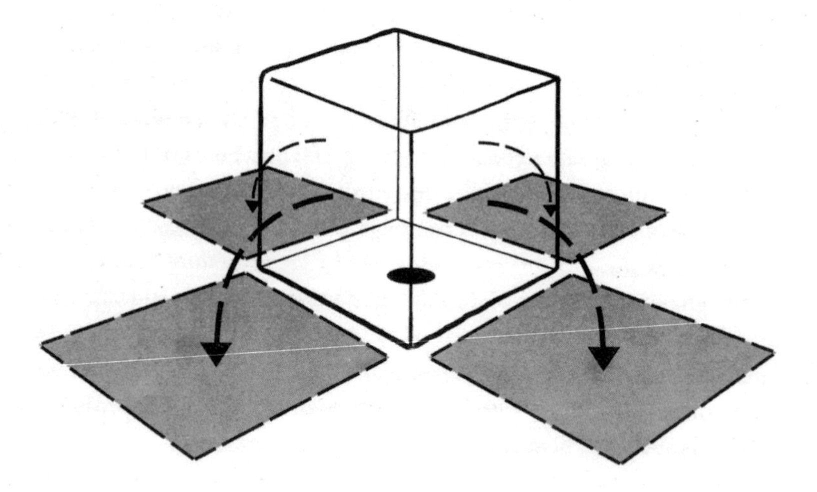

Now in your mind, roll the cube: forward, left, left, forward, right, backwards, right. Where's the dot?*

★

Too easy for you? Well, find someone to test you getting faster, and with more moves each time!

Q *How fit do you have to be to become an astronaut?*

A All astronauts have to be fit and healthy prior to going into space. That's not to say that the space agencies are looking for exceptional levels of fitness during astronaut selection. The medical screening has more to do with choosing candidates who have a long-term prognosis for good health (and therefore those who constitute a low risk of developing a medical problem whilst in space).

This week-long process during astronaut selection involved a plethora of medical tests. A strong emphasis was placed on examining cardiovascular health, eyesight and bone mineral density – all areas of the body that are greatly affected by microgravity (we will explore the full long-term medical effects of space on the human body in the 'Return to Earth' chapter).

Traditionally, about 50 per cent of candidates fail the astronaut medical screening, and this was true of our selection process. It is usually one of the last phases of the process, which often raises the question of why medical screening is not conducted sooner. However, the cost – in terms of time and money – of conducting medical examinations of a much larger pool of candidates would be prohibitive. Furthermore, some of the medical tests are invasive and not without risk, something that only a minimum number of normal, healthy individuals should be subjected to. Most unsuccessful candidates failed to meet the stringent eyesight and cardiovascular requirements. Whilst there is really very little you can do to prepare for the medical week, maintaining a healthy lifestyle will give you the best chance of success.

* Answer: Back where it started, on the bottom of the cube.

Of course, as an astronaut, it is in your best interests to be fit and healthy. After all, some of the training, such as spacewalking for several hours in a pressurised suit, is very demanding physically. Generally speaking, the fitter you are, the more you will enjoy both training and flying in space – and the easier it will be for your body to adapt to weightlessness and to readjust to gravity on returning to Earth. As an astronaut, you're expected to have at least four hours of structured physical training each week, although most people will do more than this, in addition to any personal sports. The space agencies have excellent instructors who specialise in astronaut strength, conditioning and rehabilitation. These specialists will develop a personal programme to ensure that each astronaut is fully prepared for the physical challenges of spaceflight and, more importantly, to make a full recovery in the months after landing.

Quick-fire round:

Q *My vision is not perfect. Can I still become an astronaut?*

A There is no clear answer to this question, due to the large number of visual defects that occur. The main tests that candidates need to pass include visual acuity, colour perception and 3D vision. Wearing glasses or contact lenses is not a reason for disqualification, and minor visual defects that require corrective lenses to achieve perfect vision are acceptable. However, surgical interventions (such as laser surgery) to correct visual acuity can lead to disqualification, while other surgeries are acceptable – each case is judged individually.

Q *How old was the youngest astronaut?*

A The youngest astronaut at the time of his first flight is still the Russian cosmonaut Gherman Titov. He was 25 years and 329 days old when he was launched in Vostok 2 in August 1961, as the second human in orbit.

Q *How old was the oldest astronaut?*

A The oldest astronaut to fly in space was the American astronaut John Glenn. Born in July 1921, he was 77 years old when he flew aboard the Space Shuttle for his second and last mission in October 1998.

Q *What psychological training do you do to prepare for spaceflight?*

A When I was first selected as an astronaut candidate a reporter asked me, 'How do you think you'll cope, living in a tin can for so long away from home?' It's a question that the space agencies take seriously. A long-duration mission to space can be psychologically challenging, and knowing how someone will cope begins right back at astronaut selection. One of the reasons for the in-depth psychological profiling conducted during the year-long selection process was to ensure that candidates were chosen who displayed the personality traits that are necessary for dealing with the confinement, isolation and remoteness of spaceflight.

Having selected candidates with the psychological *right stuff*, the next stage is to put astronauts in circumstances where they can learn more about themselves – and others – under challenging conditions. This started during basic astronaut training, beginning with theoretical training learning about human behaviour and performance. Knowing the theory is one thing, but there is no substitute for practical experience when learning about psychological stress and human behaviour. For this, the European Space Agency ran a survival training course for its six new recruits in Sardinia in June 2010. When I told some of my British Army colleagues that I was going to Sardinia in June to undertake survival training, understandably there was much amusement, at my expense. 'Tough call – I hear it's a bad year for Chianti' was one response. My efforts to explain that the Chianti region is in Tuscany, not Sardinia, didn't help matters. However, my point is that it doesn't take much to stress the body to a point where you begin to notice personality traits emerging that are otherwise hidden by everyday norms. Sleep deprivation, lack of food for a few days and a good dose of strenuous effort usually do the trick, whether you are in the Brecon Beacons or the Sardinian mountains.

Another excellent psychological training opportunity that the ESA provides is a caving course. This usually involves spending several days and nights living as a group of about six international astronauts in a vast cave network. The caves are technically challenging, requiring several hours of vertical ascent and descent using ropes and climbing equipment before arriving at base camp. From there, astronauts explore deeper into the caves whilst collecting microbiological samples and conducting scientific research, learning about cave photography and completing other tasks that involve a high degree of teamwork and communication skills. This provides an excellent environment that

in many ways replicates some of the stresses experienced during spaceflight and offers a valuable opportunity to learn more about yourself and others.

NASA and other space agencies provide similar training exercises to prepare astronauts for the psychological challenges of spaceflight. One of these is 'NEEMO' – NASA's Extreme Environments Mission Operations, which features later in this chapter. First, though, given the myriad of subjects that astronauts have to be trained in before flying to space, it's probably a good time to answer this question next . . .

Q *How long does it take to train to become an astronaut?*

A As a general rule, most professional astronauts will have received a minimum of three to four years of training prior to their first flight to space, although this can vary, depending on when you're assigned to a mission.

To begin with, all newly selected astronaut candidates have to go through a period of basic training, the purpose of which is to bring everyone up to a common level of knowledge across a broad spectrum of subjects. Although the approach to basic training will differ among the national space agencies that participate in the ISS partnership, there are agreed criteria regarding what this training should consist of. ESA's basic training course lasted 14 months, covering general subjects such as science, computing, orbital mechanics and space technologies. Additionally, astronauts usually start learning specific skills, such as Russian language, spacewalk training and robotic arm operations, during this time.

After completing basic training, most astronauts will probably have to wait a while before being assigned to their first mission. During this time the job of astronauts is to support the human spaceflight programme and continue to enhance their own knowledge, operational skills and professional development. This is known as the 'pre-assignment' phase, which may last for several years or just a few weeks, depending on when you are fortunate enough to be assigned to a mission. Within the European Space Agency there have been astronauts waiting up to 14 years for their first mission (Sweden's Christer Fuglesang) and others who have been assigned to a mission before even completing basic training (Italy's Luca Parmitano).

The final phase of training is called the 'assigned training flow'. This usually lasts about two and a half years, until launch. Having said that, some experienced astronauts have been assigned to a mission at very short notice.

As an example, my NASA crewmate Tim Kopra was assigned to the Space Shuttle *Discovery* mission STS-133, due to launch to the International Space Station on 24 February 2011. With a little over a month until launch, Tim was involved in a bicycle accident and unfortunately broke his hip. Furthermore, he was scheduled to be the primary spacewalker for two spacewalks during that mission. Tim was replaced by NASA astronaut Stephen Bowen and, despite the short-notice change of crew, the launch went ahead as scheduled, with Steve completing the two spacewalks. That was a pretty fast turnaround for a crew member to join a mission, but it went to prove that good training, experience and a flexible approach can overcome the most demanding of situations.

I was assigned to Expedition 46/47 to the International Space Station on 20 May 2013, exactly four years after being announced as one of ESA's new astronaut recruits. My assigned training flow lasted the standard two and a half years prior to launching to space on 15 December 2015, meaning that I had been either training or working in ISS operations for six and a half years before my mission.

Q *What are the language requirements to become an astronaut?*

A The two official languages spoken on board the International Space Station are Russian and English. But on a lighter note, there is a third language that is common among astronauts, and that is 'Runglish'. This term was coined in 2000, as the first crews were beginning to live and work on the new orbiting laboratory. Russian cosmonaut Sergei Krikalev remarked, 'We say jokingly that we communicate in "Runglish", a mixture of Russian and English languages, so that when we are short of words in one language we can use the other, because all the crew members speak both languages well.'

Learning to speak Russian is hard work. At least I found it hard work, and I know that many of my native English-speaking colleagues would agree with me. NASA astronaut and ISS commander Scott Kelly told me that it is only the first ten years of studying Russian that are difficult, and he wasn't kidding. However, it's vital to have a decent grasp of the Russian language, primarily because everything in the Soyuz spacecraft is in Russian. There is no English translation. All of our flight documentation, the instruments, the control panels, everything – it's all in Russian, and we speak to Moscow's Mission Control Centre only in Russian. No English is spoken inside the Soyuz, unless it's an informal conversation between the crew.

For that reason alone we need to be competent at understanding Russian and making ourselves understood. No one is looking for grammatical perfection or an extensive vocabulary. However, in order to fly to space we need to pass the American Council on the Teaching of Foreign Languages (ACTFL) oral-proficiency interview in Russian at the 'Intermediate High' level. In addition to the operational reasons for speaking Russian, there is also a social necessity. We spend many months training in Russia, engaging in cultural and social events with our cosmonaut colleagues, so it helps to be proficient in their language.

Russian language training begins almost immediately for all

astronauts as soon as they start basic training. The European Space Agency engaged the assistance of a Russian-language school in Bochum, Germany. Three of the first six months of our basic training consisted of Russian language training, to include a month-long immersion course in St Petersburg, living with a Russian family.

The intensity of these first few months of learning Russian prepares you well for your initial visit to the Gagarin Cosmonaut Training Centre, otherwise known as 'Star City', where you learn all about the Russian segment of the ISS and the Soyuz spacecraft. But a short, sharp burst of language training doesn't build a strong foundation. It took regular lessons over a period of two to three years before I began to feel more comfortable speaking the language and leant less heavily on our Russian interpreters during the many hours of technical training and simulations at Star City.

Q *Did you train in the centrifuge, and did it make you feel sick?*

A All astronauts who fly to space in the Soyuz complete training in the centrifuge at Star City, Russia. The thought of being strapped into a confined space and being spun around until you feel eight times your own body weight may seem terrifying – perhaps conjuring up images of James Bond's cheeks being deformed, and of him nearly passing out, as he tried to stop the centrifuge in the film *Moonraker.* However, our instructors did a great job of preparing us for what the centrifuge would feel like and, more importantly, how to breathe correctly. I actually found it a huge amount of fun and would gladly repeat the training, if given the opportunity.

Despite spinning constantly, the centrifuge does not make you feel dizzy or sick. This is because the capsule that you are lying in is free to pivot on a hinge, which means that as it gets faster, the force of acceleration (g-force) is always felt in the same direction, from the front to the back of your chest. We adopted the same, almost foetal, sitting position that the Soyuz seat demands, which meant we experienced very similar g-force to how it would feel during launch and re-entry. Inside

the centrifuge you don't actually get a sensation of spinning, but instead it feels as if you are constantly accelerating in a straight line.

Good breathing technique is really important. The highest stress that we have to endure during centrifuge training builds up to eight times the force of gravity (8g), where it remains for 30 seconds before reducing again. This replicates a 'ballistic re-entry' – when something goes wrong with the spacecraft (I talk about the perils of ballistic re-entry later in the book). As the g-load gradually builds, it feels as though someone is placing weights on your chest. Not only does this make it harder to breathe, but you also feel as though you have to tense your muscles to prevent your chest from collapsing in on itself. Actually this is the best advice for withstanding g-force – tense your muscles and try and 'lock' your chest in place, whilst breathing using your stomach, in a gulping fashion. We practised this breathing technique at 4g before increasing it to 8g.

Tolerating 8g for 30 seconds was actually quite hard work. The best way to describe it is like doing a weight-training exercise. Imagine bench-pressing your body weight for 30 seconds. It may seem okay at first, but as time goes on it gets harder. I was surprised by how easy 8g felt for the first ten seconds, and equally surprised by how much harder it felt for the final ten seconds. It was another good reminder of why it's important for astronauts to be in excellent physical shape before flying to space.

Q *How do you train for weightlessness, here on Earth?*

A There are a couple of ways in which astronauts can train for weightlessness without actually going into space. One method is to use the neutral buoyancy of water to simulate weightlessness, which I talk about in more detail in the 'Spacewalking' chapter. However, the best way to train for weightlessness is, unsurprisingly, to become weightless! In Earth's orbit astronauts are essentially in freefall when they are inside a spacecraft that is also falling to Earth. We can simulate this, without leaving the planet, by floating inside an aircraft that is falling to Earth – although the difference is that at some point the aircraft has to pull out of the dive to avoid crashing!

We call this a parabolic flight or, to give it its nickname, the 'vomit comet'. In order to achieve weightlessness, the pilot must first pull up into a 45-degree nose-up climb. Then, by pushing forward on the controls, the pilot can accurately target 'zero-g' all the way until the aircraft is pointing about 45 degrees nose-down, at which point they can pull out of the dive and, after a short reprieve, repeat the manoeuvre. By targeting zero-g, the aircraft actually flies part of a parabola (a symmetrical curve), hence the name 'parabolic flight'. Each parabola provides about 25

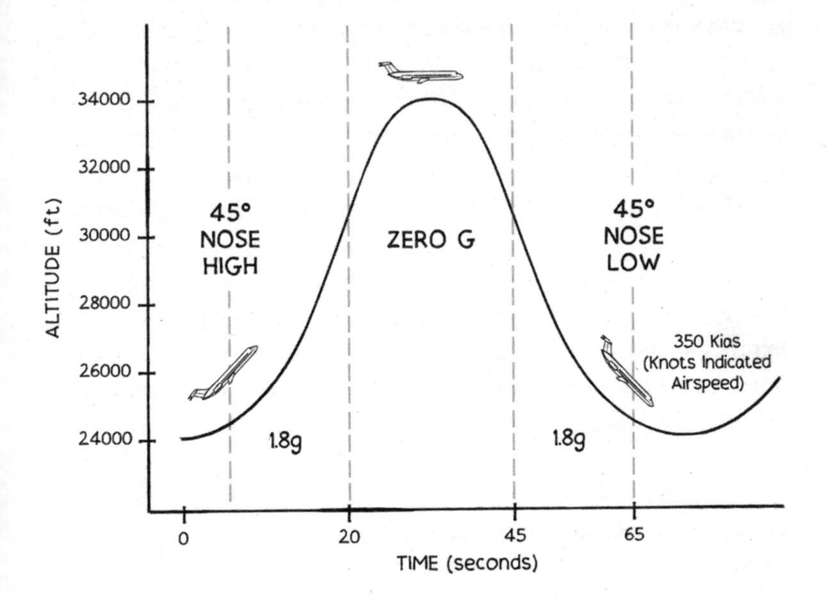

seconds of weightlessness, during which time astronauts can experience what it will feel like in space and practise some basic tasks, such as handling objects, controlling the body and even running on a treadmill or exercising. In addition to astronaut training, parabolic flights are also used extensively for scientific experiments that don't require extended periods of time in microgravity.

There are several places in the world that offer this type of training, with flights usually lasting around three to four hours and incorporating between 30 and 60 parabolas, so it's not for the faint-hearted. Up to two-thirds of participants might end up feeling unwell, hence the unfortunate nickname. However, anxiety has a lot to do with this and, with good training and perhaps a dose of mild anti-nausea medication, there is no reason why most people shouldn't thoroughly enjoy the experience. My first parabolic flight was during basic astronaut training and I remember it being a huge amount of fun. That initial feeling of lifting off the floor of the aircraft, floating in the cabin, was incredible and brought big grins to my face and those of my five classmates.

Q *What do astronauts do when not in space?*

A Following completion of basic astronaut training, my first task was to head to the ISS Mission Control Centre in Munich and qualify as a Eurocom. The role of a Eurocom is to be the person who talks to the astronauts on the space station. You are the voice link between the ISS and the ground teams, responsible for consolidating all the information that is coming in from a myriad of sources on the ground and putting it into clear, concise instructions to the crew. It's one of the most valuable jobs in preparing you for what life on board the space station is really like. This training took several weeks, but on completion I was proud to be a part of the close-knit Mission Control team that supports daily operations on board the ISS.

During this 'pre-assignment' phase I was also fortunate to be the first European astronaut selected for a NEEMO mission with NASA. NEEMO, 'NASA's Extreme Environments Mission Operations', involves working with an extraordinary group of scientists, engineers and other astronauts. I like to think of NEEMO as the Special Forces of NASA's human spaceflight team – a small group of innovative experts who are developing the technologies and systems for future space-exploration missions. I was assigned to the 16th NEEMO mission, with the objective of developing the tools, techniques and procedures that would be required for a future mission to an asteroid.

The 'extreme environment' part of NEEMO lends itself to the fact that these missions are conducted underwater – about 20 metres underwater in fact. To achieve our objective, I and five other crew members planned to live underwater for 12 days, conducting diving expeditions each day from the sanctuary of our marine habitat, 'Aquarius'. This research station lies nearly 6 km off Florida's Key Largo coastline, anchored to the ocean floor, and is ideally suited to simulate space missions, for several reasons. First, it's a very confined environment, with just enough room to support six crew members in fairly cramped conditions – it makes the ISS look positively luxurious.

Second, there's real risk involved. When we breathe air underwater, the increased pressure causes nitrogen to be dissolved into our bloodstream and transported to our body tissues. Do this for any length of time and our body tissues become saturated with more nitrogen than they would otherwise have on the surface. In a sense, a body saturated with nitrogen is a bit like a shaken-up fizzy drink. There's a lot of dissolved gas trapped in there, but everything is fine . . . so long as no one takes the cap off! When diving, coming to the surface rapidly is akin to taking the cap off. A swift reduction in ambient pressure releases the excess gas in an explosion of bubbles. This is fun to do with a fizzy drink, but not so much fun in the human body. As bubbles are released into body tissue and the bloodstream, symptoms can vary from itching and pain to paralysis and even death.

And so, during our 12-day mission underwater, we did not have the option of a quick return to the surface if something went wrong. In fact, in order to release this dissolved gas safely we needed to gradually 'decompress' over a period of 18 hours – not helpful if there was a fire in the habitat, or a drowning incident during one of our dives. This element of risk is something that all astronauts must deal with during their career. The space station is an environment where decisions have real consequences for the safety of the crew, and training exercises such as NEEMO provide excellent simulation of a space mission.

However, the most important reason for going underwater was to use the neutral buoyancy that water offers to simulate weightlessness. Whilst we can use a swimming pool for some of this research (and frequently do), the ocean allows us to experiment with larger pieces of equipment, such as deep-water submersibles, which can simulate space-exploration vehicles.

Q *What subjects do you have to study during mission training?*

A It might be easier to answer this question by stating what subjects we don't have to study during training! Astronauts receive an awful lot of training. In fact, one of the hardest things to figure out, as a rookie astronaut, is what stuff it is really important to retain when the memory bank gets a bit full. As it happens, the training programme and our dedicated, ever-patient instructors helped us with this, because as we got closer to launch, the really important subjects were repeated over and over again.

Some of the main areas that astronauts focus their attention on are extravehicular activity (EVA, or spacewalking), use of the space station's robotic arm to capture visiting cargo vehicles, emergency training and Soyuz spacecraft training. This is not to say that the hours and hours spent studying other topics are unimportant, but something catastrophic is less likely to happen if a mistake is made with them.

In addition to the subjects already mentioned, astronauts have to learn about the space station in some detail, to include all of the life-support systems, electrical systems, thermal control systems, guidance, navigation and control systems, to name just a few. Furthermore, astronauts will spend the majority of their time conducting scientific research, and therefore they must learn about the different laboratory equipment, or 'payloads', on board the ISS. Then there are areas such as medical training, language training and survival training that must be completed. Being an ISS crew member is a bit like being a 'Jack of all trades' . . . and hopefully master of some!

Q *Do all astronauts receive the same level of training?*

A Generally speaking, astronauts receive the same basic level of training that enables them to be an effective crew member on the ISS. This is important in order to have the maximum flexibility for space-station operations. It's useful if all astronauts are able to work effectively in each of the different science laboratories on the ISS, can conduct basic maintenance tasks and have the ability to perform a spacewalk or use the space station's robotic arm.

That said, there are some important differences. The ISS is the most complex structure that humans have assembled, and so when it comes to maintaining the space station, astronauts have specific roles and receive different levels of training in order to distribute this mammoth task.

Astronauts are assigned as a User, Operator or Specialist in various segments of the ISS. For example, I was a Specialist in the European laboratory (*Columbus*) and Japanese laboratory (*Kibo*), an Operator in the US laboratory (*Destiny*) and a User in the Russian segment. That meant that, as a Specialist, I was the 'go-to' guy for any serious maintenance activity required in *Columbus* or *Kibo*. As an Operator, I could perform routine maintenance tasks in *Destiny* and the other US modules, and could assist a Specialist in some of the more complex tasks. As a User in the Russian segment I had been trained on all the systems and could operate them safely and effectively, but under normal circumstances I would not be expected to conduct maintenance activities.

There are several other factors that determine the level of training required, such as which vehicle you're flying to space in. At the time of publication, the only spacecraft taking crew to and from the International Space Station is the Russian Soyuz spacecraft. However, the Space Shuttle visited the ISS 37 times, and some astronauts are already training to fly on commercial spacecraft, such as Space-X's Dragon or Boeing's CST-100, which will launch to the ISS in the near future. Each vehicle has unique training requirements, and on some spacecraft crew members will have different levels of training, depending on whether they are the pilot, commander or flight engineer, and so on.

The Soyuz has three seats, with the commander (until now always a Russian) occupying the centre seat. This is the only position from which you can actually fly the spacecraft, using two hand-controllers, if manual control is ever required. Under normal circumstances the crew interact and send commands to the Soyuz, but do not need to fly it manually. The left-hand seat (Flight Engineer 1) is the backup commander. That crew member has received a level of training that would enable him or her to fly the Soyuz, in the event that the commander was incapacitated. The right-hand seat (Flight Engineer 2) can vary from receiving a bare minimum of training, in order to take care of personal life-support systems, to being a backup left-seater.

I flew in the right-hand seat, but was fortunate to receive a very comprehensive training package on the Soyuz. Furthermore, our commander (Yuri Malenchenko), having flown five times previously, was so experienced that in the early stages of training he would sometimes leave Tim Kopra and me to do the simulations ourselves. Tim would jump into the commander's seat and I would fly the left-hand seat. Having the ability to understand other crew members' roles in a spacecraft makes a huge difference to how well that crew can function together.

Q *What was the worst part of training?*

A Well, first I should say that you have to keep things in perspective – so when you're having a bad day as an astronaut who is training for a mission to space . . . that's still a pretty good day! That said, during training there were a few things that I didn't always look forward to. I've already mentioned that learning Russian was a struggle and did not come naturally to me. When I first started Russian lessons I was being taught by a German who spoke Russian and English, and by a Ukrainian who spoke German, but not so much English. Both instructors were excellent and the lessons were a lot of fun, but I couldn't help wondering if something was being lost in translation. Russian grammar, with its six cases obeying grammatical gender, seemed a complete mystery to me.

Our training took us all over the world and, wherever we spent any length of time in one place, we had Russian lessons. That meant I was exposed to several different Russian tutors in Germany, the USA and Russia. It soon became apparent to me that nothing was being lost in translation – it was simply a case of a lost student! I began to recognise that familiar expression on my tutors' faces as they pondered how somebody could so easily fathom the inner workings of a space station, but hadn't a clue when to use the genitive or dative case. I persevered and achieved the required standard, but if the truth be known, those long hours of Russian lessons and late nights studying grammar were not among my happiest memories of training.

Another fairly unpleasant memory that I have comes from my NEEMO mission. Now NEEMO ranks as one of the top ten experiences I have ever had in my life, so why does it deserve a mention as one of the worst parts of training? Well, like many potential 'aquanauts' embarking on a NEEMO mission, I had expected our underwater habitat to feature some sort of lavatory – a chemical latrine perhaps. It does, but the loo was there only for use at the very end of the mission, during the 18 hours of decompression. Before that, it was out of bounds. There was no system for flushing waste out of the habitat, and so the aroma would rapidly become extremely antisocial, with six people using the latrine for 12 days. To that end, when nature's duty called, we did what the fish do and used the ocean. When in Rome . . .

I don't have a problem with peeing in the ocean, and in Aquarius it was as simple as a trip to the wet porch to relieve oneself, although politeness dictated that you checked beforehand that no one was about to surface from a diving trip. However, when burdened with the other call of nature, we would use the 'Gazeebo'. This was like a large upturned bath with an air pocket, capable of accommodating six people at a squeeze and located just a short duck-dive away from Aquarius. Its primary purpose was for use in the event of an emergency. Due to the fact that there was no safe return to the surface without first decompressing, the crew needed somewhere to shelter in the event of a fire or flood in Aquarius. More routinely, the Gazeebo fulfilled a secondary purpose – as our makeshift loo, being a comfortable distance away from Aquarius.

Without going into too much detail, trips to the Gazeebo were frequently eventful and seldom pleasant. Unfortunately, human waste was *haute cuisine* to a whole host of fish that lived in the vicinity of Aquarius. Just the mere presence of a human standing in the Gazeebo would be the rallying cry for an army of tropical marine life to form ranks and prepare to attack. Triggerfish were the worst. They have jaws and teeth designed for crushing shells and are notoriously ill-tempered. Aquanauts frequently returned from the Gazeebo dripping blood from a finger or bum-cheek and, whilst nursing these wounds, we would sympathise with one another and discuss the merits of various techniques for keeping these aggressors at bay.

This became such a problem that it was decided to fit the Gazeebo with 'The Bubbler' – a makeshift length of tubing pierced with holes

and connected to an air supply. The idea was that turning on the air supply would create a wall of bubbles, within which you could stand and do your duty. This worked for about 48 hours, after which the fish wised up and The Bubbler simply became a call to supper. I wouldn't say I was scarred by the experience, but it was definitely not a highlight of my training. So, from the worst part of training, let's now take a look at . . .

Q *What was the best part of training?*

A I thoroughly enjoyed the vast majority of my training and there were many highlights: parabolic flights, NEEMO, caving, survival training, to name a few. In particular I loved everything to do with training for a spacewalk. The spacesuit itself is a mini space station, designed to keep you alive in the harsh vacuum of space for eight hours or more, and it is a phenomenal piece of engineering in its own right. This is partly achieved by a 'Portable Life Support System' (PLSS) or backpack, which also incorporates a small jet-pack. Called a 'Simplified Aid for EVA Rescue', or SAFER, this relies on small nitrogen-jet thrusters that allow an astronaut to manoeuvre in space, in case they become detached from the space station during a spacewalk.

The 24 high-pressure thrusters enable control in six axes – pitch, roll, yaw, forwards/backwards, sideways and up/down (we call those last three the x, y and z axes). The astronaut controls the thrusters using a single hand-controller – and if you think that sounds like something from an early James Bond film, then you wouldn't be far wrong. SAFER offers a stranded astronaut a last-ditch attempt to reach the sanctuary of the space station. However, unlike an earlier jet-pack (called a Manned Manoeuvring Unit) that contained enough propellant for a six-hour EVA depending on the amount of manoeuvring done, with SAFER there's really only enough fuel for one self-rescue attempt. Nothing like a bit of pressure to add to an already tricky situation!

So how do astronauts learn to fly a jet-pack? The answer lies at NASA's Virtual Reality Laboratory at the Johnson Space Center in Houston. This fully-immersive training facility provides an incredibly realistic environment where astronauts are repeatedly sent tumbling off into space, until they have mastered the correct technique to get safely back to the space station. The trick is first to stop the tumbling motion and then find the space station. If you're lucky, you might be able to see the ISS already, or another reference point, such as Earth. If not, then

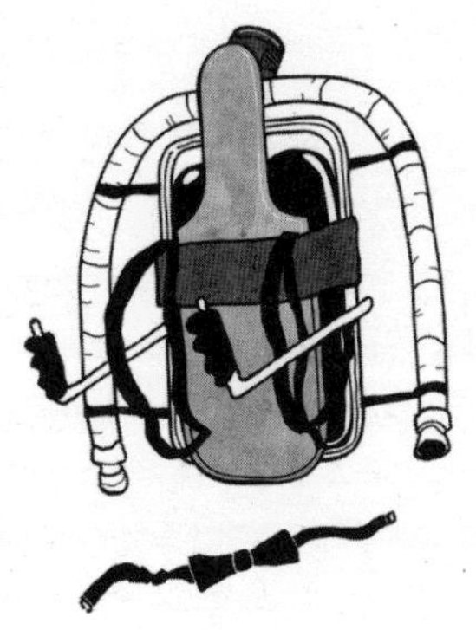

valuable fuel has to be expended searching for the space station. Once you've found the ISS, you will still be drifting away from it, and so time is of the essence. The further away you get, the more other factors (such as the rules of orbital mechanics) come into play that make the task of getting back that much harder. The most important part is to aim accurately at the point where you came off the space station, before thrusting forward. Get that wrong – and you'll end up fighting your course the whole way back and running out of fuel.

Astronauts practise being thrown off-station around 20 to 30 times (including at night) before taking a final exam. However, self-preservation provides all the incentive necessary to pass that particular test.

Quick-fire round:

Q *What's the best advice you've received? – Alex Gellersen*

A A teacher of mine used to say, 'Life is like a dustbin – you get out what you put in.' This may seem like rather harsh advice to give a teenager, but it has served me well over the years and still does. I never expect anything to come my way without hard work, patience and determination.

Q *Is it true that astronauts learn to sweat the small stuff?*

A Absolutely. I learnt that, as a test pilot. When the big stuff goes wrong, knowing the small stuff will give you options.

1: Although the most physically challenging, EVA training was also, for me, the most enjoyable part of astronaut training.

2: This photo was taken before NASA had an approved UK flag for the suit!

3: My helicopter training was incredibly useful when learning to control the ISS's robotic arm. It requires a high degree of coordination and spatial awareness, with each hand controlling motion in different axes. (Training at NASA JSC in the Cupola simulator with crewmate Tim Kopra.)

4: A parabolic flight aboard the Airbus A300 Zero-G as part of my training – experiencing weightlessness takes a lot of getting used to.

5: Vital survival training ahead of our NEEMO mission underwater. Here, we are practising initiating rescue breaths through the pocket mask in an attempt to resuscitate an unconscious diver.

6: Another training exercise for spacewalking. This Partial Gravity Simulator (POGO) suspends astronauts in a harness, which allows them to understand how their body will react to torque inputs when using tools. A firm, stable body position is vital for EVA tasks. It's also a lot of fun!

7: The return to Earth can be punishing on the body. It's important to have an individually tailored Soyuz seat to protect us from the rigours of re-entry, parachute opening and landing. A bath of gypsum a few months prior to launch ensures a perfect fit!

8: Some of the delights on offer at the space station café. Most food is either irradiated, dehydrated, canned or 'off-the-shelf' dry goods. Space food is generally fairly bland...

… although we did have a little help from Heston Blumenthal and the Great British Space Dinner (photo 9). The canned Alaskan salmon was my favourite ISS dinner, as the intense flavours of the capers worked really well in space.

10: Winter survival training in the -24°C temperatures of Moscow in January helped us to be prepared for anything.

11: Although we avoid doing laundry up in space, there was no excuse during training…

12: If a fire were to break out on the ISS, it's vital to locate and contain it as quickly as possible, so we practise this repeatedly during our training.

13: Much of the science we do on the ISS helps us to better understand the human body. We collect medical data before, during and after a space mission to identify what changes have occurred. This experiment monitored airway inflammation at ambient and reduced pressure, providing valuable research that will benefit asthma sufferers.

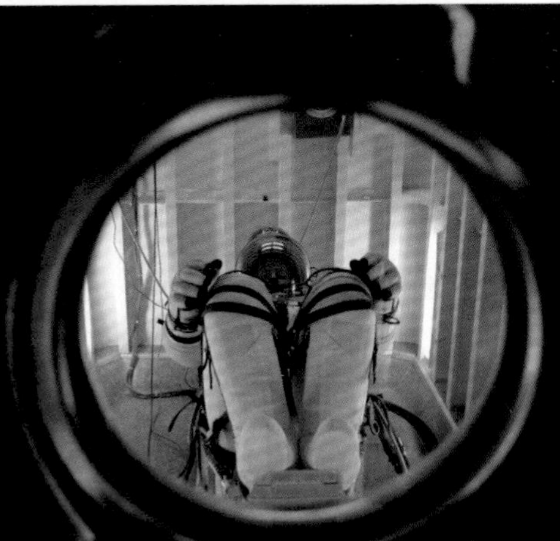

14: Vacuum chambers mimic the conditions of space so that we can test our spacesuits.

15: Last practice run in the Soyuz simulator before launch day!

16

17

18

There are various traditions leading up to launch day including: planting a tree in the Avenue of the Cosmonauts grove in Baikonur (photo 16), signing of the crew-room door (photo 17) and blessing of the crew by a Russian Orthodox priest (photo 18).

19: One final farewell to my family from behind the glass before heading to the launch site.

20 and 21: 15 December 2015, Launch Day.

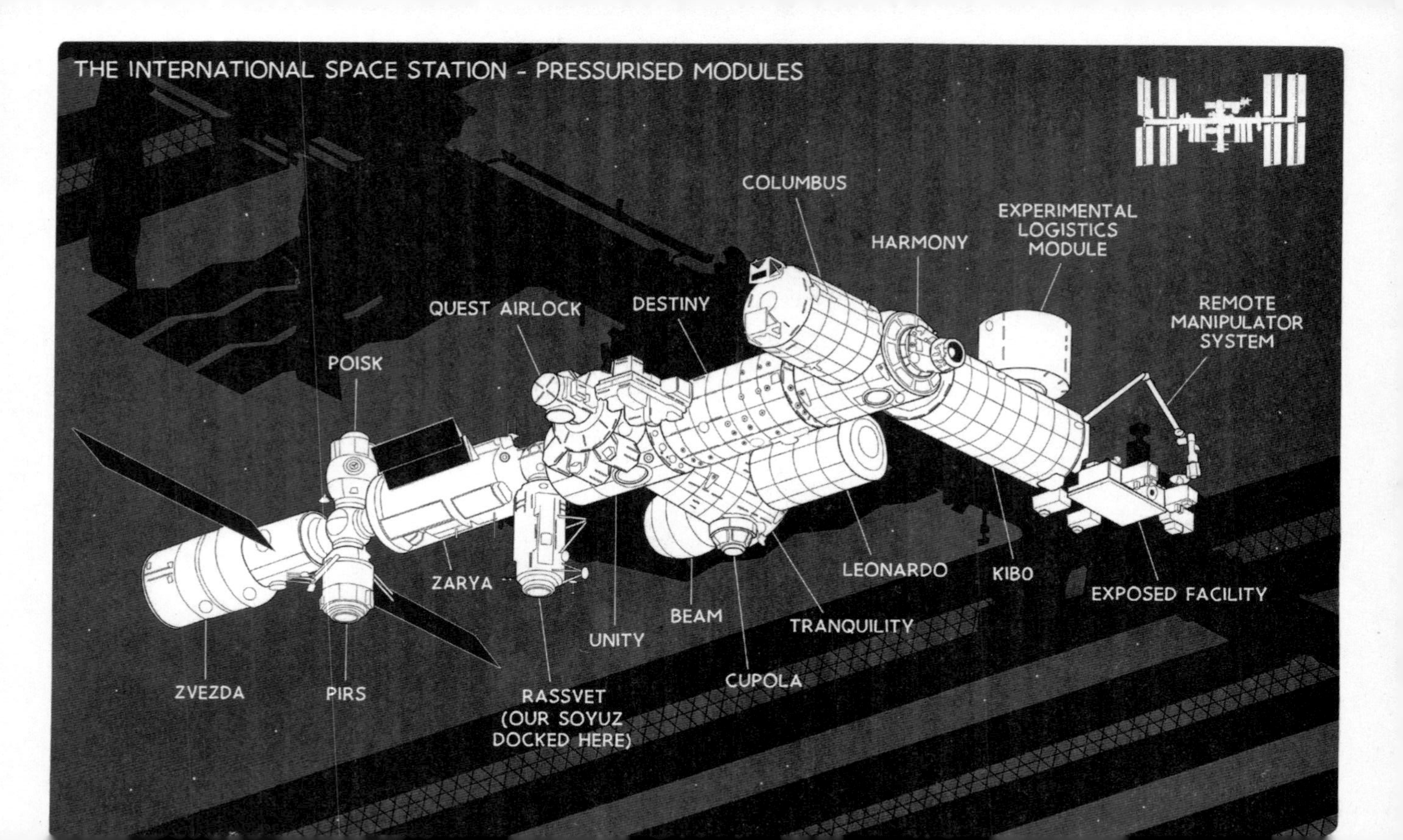
THE INTERNATIONAL SPACE STATION - PRESSURISED MODULES
COLUMBUS
HARMONY
EXPERIMENTAL LOGISTICS MODULE
QUEST AIRLOCK
DESTINY
REMOTE MANIPULATOR SYSTEM
POISK
ZARYA
LEONARDO
KIBO
EXPOSED FACILITY
BEAM
TRANQUILITY
UNITY
ZVEZDA
PIRS
RASSVET (OUR SOYUZ DOCKED HERE)
CUPOLA

LIFE AND WORK ON THE ISS

Q *What's a typical day like on board the International Space Station?*

A From the moment I arrived, every day on board the ISS was an exciting, challenging and stimulating experience. Every nook and cranny of the space station reveals a mini-miracle of science. In the spaces where experiments are not installed, a myriad of complex systems churn away perpetually, to provide something that many of us take for granted here on Earth – clean air and water. At first the ISS can seem a daunting place to work, harbouring a wealth of intricate equipment, a multitude of computers (52 at the last count) and more than 12 km of electrical wiring! And that's before you throw in the challenge of living in microgravity. But for an astronaut, first and foremost the ISS is your home as well as your office, for months at a time. Okay, you might float in on the morning commute, and the view of Earth from the Cupola window is breathtaking, but the ISS is primarily a place where you work hard, perform experiments, eat, sleep, exercise and spend time with colleagues.

As such, life on board quickly falls into a routine and becomes surprisingly normal, even though only a few millimetres of aluminium protect you from the vacuum of space. I should make clear: life on board is never boring for a moment. But this 'normalisation' is an essential process. In order to function efficiently and effectively as a crew member on the ISS, you cannot be constantly distracted by the awe and wonder of circling the planet 16 times a day whilst travelling at ten times the speed of a bullet – as thrilling as that is. And whereas during launch, or on a spacewalk, astronauts maintain a constant vigilance against the possibility of something going wrong, inside the space station is a place where you can relax and feel relatively 'safe'.

Before I answer your questions about everyday life in space, let's first examine the extraordinary feat of engineering that cradles life beyond the sanctuary of our planet: the International Space Station.

Q *What exactly is the International Space Station?*

A The quick answer:

1 The largest and most sophisticated spacecraft in history.
2 A cutting-edge science laboratory.
3 A home in space for astronauts.

A The slightly longer answer: the ISS is the most advanced human-made structure ever built. Weighing more than 400 tonnes and covering an area as big as a football pitch, the ISS orbits Earth approximately 400 km above the surface, travelling at a speed of 27,600 km/h. This means it circumnavigates the globe every 90 minutes.

If I was an estate agent, I would say that the ISS is as big as a six-bedroom house. It boasts two bathrooms (although no shower, I'm afraid), a gym and a large 360-degree bay window, known as the Cupola.

If you're looking for pressurised space, you're in luck. It has more than 820 cubic metres (about the same as a Boeing 747-400 'Jumbo Jet'), more than enough room for its crew of six people and a vast array of scientific experiments. The price? It is estimated to have cost more than $100 billion to construct, which makes it a likely candidate for the most expensive single item ever built.

A The long answer: the ISS was built in partnership by five different space agencies representing 15 countries: NASA, Russia's State Corporation for

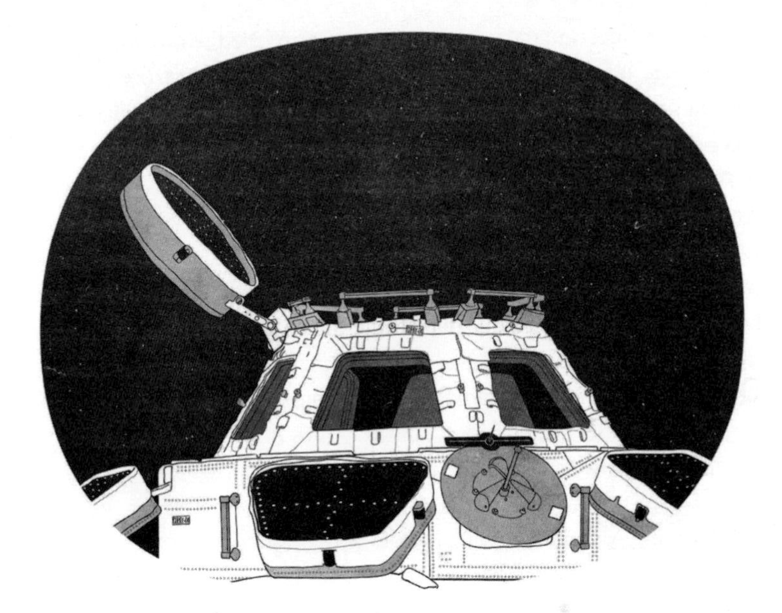

Space Activities (Roscosmos), the European Space Agency, the Canadian Space Agency and the Japanese Aerospace Exploration Agency. The sheer size and weight of the ISS prevented it from being assembled on Earth – there's simply no rocket large enough that could have carried it to space, for one thing. Instead, the ISS was assembled in space, like some giant Meccano or Lego construction set, with the majority of the modules delivered in stages over a 12-year period. The ISS is divided into two segments, the Russian Orbital Segment and the United States Orbital Segment (which is also shared by Europe, Canada and Japan).

ISS construction began in November 1998, with the launch of Russia's *Zarya* module (meaning 'dawn'), followed two weeks later by the US *Unity* module. It has been permanently occupied since November 2000, when NASA astronaut Bill Shepherd and Russian cosmonauts Yuri Gidzenko and Sergey Krikalev arrived in a Soyuz spacecraft on Expedition 1. Assembly was delayed for two and a half years due to the tragic loss of the Space Shuttle *Columbia* in 2003. It's hard to say when ISS construction was completed, since it is constantly evolving. At the time of writing, the latest module to be installed was the Bigelow Expandable Activity Module (BEAM), delivered by SpaceX's CRS-8 Dragon spacecraft in April 2016. I had the honour of capturing that spacecraft, using the space station's robotic arm.

As of 2017, it has taken 32 rocket launches from the Space Shuttle (27), Proton (2), Soyuz-U (2) and Falcon 9 (1) to deliver the ISS structure and modules into low-Earth orbit, and more than 140 launches to provide crews, logistics, servicing and resupply missions. Furthermore, more than 1,200 hours of spacewalking have been required to assemble and maintain the ISS in orbit. It's hard to overstate the scale and complexity of constructing and operating the ISS. I once heard a NASA engineer describe its construction as like dropping all the pieces of a container ship in the middle of the Pacific Ocean and then trying to build it at sea.

Q *What are all the different parts of the space station?*

A The space station is essentially made up of a number of pressurised modules that serve as scientific laboratories, docking ports, airlocks, storage and habitation areas. Then there is a 'truss' that forms the spine of the space station. The truss consists of 12 metal lattice structures that span the 109-metre width of the space station, containing many of the components that keep the ISS alive, such as power, cooling, communications, attitude and control systems, and more. The truss also houses a number of logistics platforms where many spare components have been pre-stored. Attached to the aft side of the truss are radiators (waffle-shaped panels with ammonia flowing through them, to get rid of excess heat that builds up in the station).

At each end of the truss are giant solar arrays, which turn sunlight into electricity. The truss has a joint that enables these solar panels to rotate 360 degrees to track the Sun. These wide, flat solar panels cover a total area of approximately 2,500 square metres (enough to cover eight basketball courts) and produce up to 120 kilowatts of electricity, sufficient power for 40 average homes. During darkness, batteries stored in the truss, which are charged during daylight, provide power to the ISS.

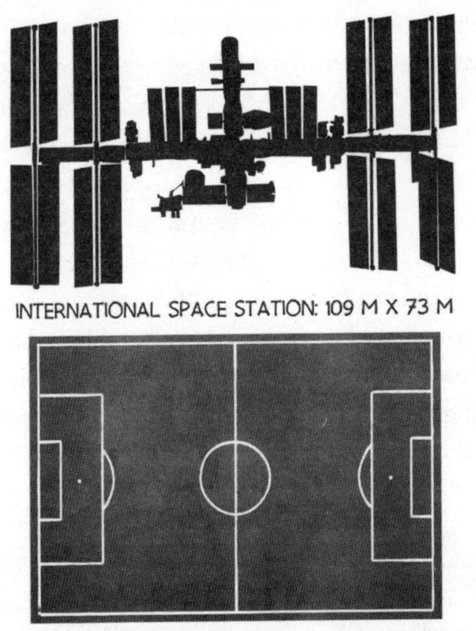

A robotic arm, known as the Space Station Remote Manipulator System (SSRMS), was constructed by Canada and helped to build the ISS by grappling and moving modules, or by moving astronauts during spacewalks to work on various parts of the space station. Either end of the robotic arm can serve as an anchor point, which means it can 'walk' around various parts of the structure, depending on where it is needed. One of the most vital tasks for the robotic arm is capturing and berthing visiting cargo vehicles that cannot dock automatically to the ISS.

In addition to the US laboratory *Destiny*, there is a European laboratory *Columbus* and a Japanese laboratory *Kibo*. There are also external platforms dedicated to research in the harsh environment of space, and experiments such as the AMS-02 or Alpha Magnetic Spectrometer (great name!) mounted on the truss. AMS-02 is a particle-physics experiment designed to measure antimatter in cosmic rays and to search for evidence of dark matter. The Russians plan to launch a laboratory called *Nauka*, along with a European robotic arm, later this decade.

THE ISS ROBOTIC ARM CAN 'WALK' AROUND THE SPACE STATION USING EITHER END AS AN ANCHOR POINT.

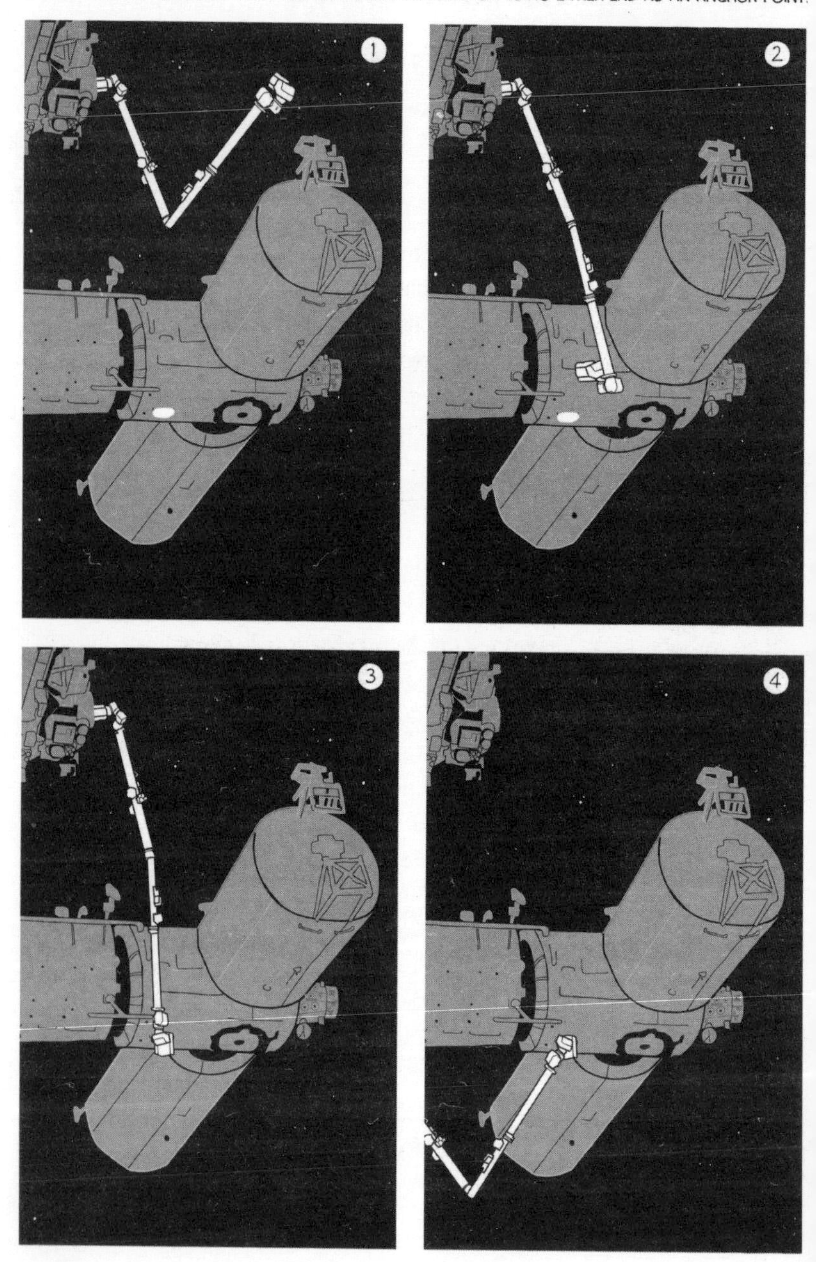

In total, the ISS is built from more than 100 main pieces. The diagram on page 78 shows all of the pressurised modules of the space station.

Did you know?

- It is believed that approximately 20 per cent of the mass of the universe is made up of 'dark matter', a mysterious ingredient that scientists cannot actually 'see', because it does not emit light or energy. Although we haven't detected it yet, we know that dark matter is there because the orbits of stars and rotations of galaxies are being driven by gravitational effects far greater than should be possible, given the mass of visible matter alone. Dark matter provides the answer to this missing mass throughout the universe. Collisions between particles of dark matter should be creating an excess of charged particles, and the AMS-02 experiment on the ISS is helping us to better understand this strange phenomenon by detecting these particle emissions.

- Antimatter is another enigmatic building block of the universe. It is the opposite of a normal particle of matter, and has a different electrical charge. For instance, the antiparticle of an electron is known as a positron. Experiments on the ISS and at the Large Hadron Collider at CERN demonstrate that when antimatter particles combine with particles of matter, they destroy each other and produce energy. This has led some scientists to believe that antimatter could one day be used as an efficient source of fuel to power spacecraft, but we are currently a long way off that scenario – at present, it requires far more energy (and cost) to create antimatter than the energy we can produce from an antimatter reaction.

Q *But what's the point? – Jeremy Paxman,* Newsnight

A Okay, so Mr Paxman didn't actually respond to my request for contributions to this book, but he did ask me this very question during an interview on *Newsnight* in 2013. There are many ways to answer it. Simply put, I see the purpose of the ISS as twofold:

- To further scientific knowledge and understanding, for the benefit of people on Earth.
- To facilitate human exploration beyond our planet.

The reason the ISS is an attractive place to conduct scientific research is that the conditions in space are different from those on Earth – and when you change the conditions (or parameters) of an experiment, you can investigate results and learn new things. Gravity is the only environmental parameter that has remained constant during the period of the evolution of living matter on Earth. The ability to change this and study the long-term effects of weightlessness – not only on living organisms, but on a multitude of physical and chemical processes – has established the ISS as a platform for groundbreaking scientific research. There have been more than 1,200 journal publications showcasing the findings of ISS research, which have increased our knowledge, pushed technology to new limits, improved human healthcare on Earth and benefited our environment.

However, mankind's outpost in space is not just about scientific research. It is also an extension of an innate human trait: the desire to explore. We are a curious species, with a lifelong capacity to learn. But in order to learn we need to explore, to test, to try new things and occasionally to leave the beaten track. By adding a healthy dash of curiosity into our genetic make-up, evolution is refining us into the ultimate learning machine. This primal urge, which has driven us to embark on voyages of discovery since *Homo erectus* first ventured beyond Africa two million years ago, is certain to pay dividends in the future. After all, if we remain a single-planet species, then the unavoidable truth is that we are headed for extinction.

Did you know?

- In 24 hours the space station travels approximately the distance it takes to fly from Earth to the Moon and back.
- At night, the ISS can be seen from Earth with the naked eye. It appears as a bright white dot, due to reflected sunlight, rivalling the planet Venus for its brightness in the night sky. It takes about ten minutes to move across the sky from one horizon to another, although it's not usually visible for this long, as it will pass into the shadow of Earth. It is often mistaken for an aeroplane, but you can tell them apart as the ISS has no flashing strobe-lights. Want to try spotting it? You can use the ISS tracker to know where to look: http://www.isstracker.com.
- The ISS is the ninth manned space station, following the Soviet and later Russian Salyut, Almaz and Mir stations, as well as Skylab from the US. Occupied since November 2000, the ISS boasts the longest continuous human habitation in low Earth orbit, having surpassed the previous record of nine years and 357 days held by Mir.
- Approximately 7 tonnes of supplies are required to support a crew of three for about six months on the ISS.
- With the arrival of Expedition 53 in September 2017, 228 different people have visited the ISS. I was number 221. My Soyuz commander, Yuri Malenchenko, is currently the only person to have visited the ISS five times.

Q *What was the first thing you did when you arrived on the ISS?*

A When Tim, Yuri and I arrived on board the ISS we were running late! Our eventful docking had put us about 30 minutes behind schedule, and when the hatch finally opened there was no time to waste. Traditionally there is a welcome ceremony when a new crew arrives on the space station. This takes the form of a video-conference between the crew and a small group of family, friends and guests assembled in Baikonur. Usually the crew would have had time for a quick trip to the loo, and maybe even a bite to eat, before commencing the welcome ceremony. In our case, we were ushered straight from our Soyuz into the Russian Service Module (*Zvezda*) to gather for the call, since the communications window that enabled video-streaming was rapidly diminishing. Then, following our first live video link-up with the ground, we finally had the opportunity to get changed, start unpacking and become acquainted with one of the most vital pieces of space hardware – the ISS loo.

Q *How do you go to the toilet in space?*

A This is by far the most popular question I get from younger children. Okay – here it is! Going to the loo in space is not so different from doing your business on Earth, but there are a few important things to remember. First, we do enjoy some privacy, since the loo is screened off in an area about the size of a telephone booth. Inside, there are some foot restraints that we use to keep ourselves stable (the fewer things floating around, the better). We pee into a hose that has a conical-shaped receptacle with a switch on the side. The most important thing to remember is to first turn the switch, which operates a fan. The whole concept of the space-loo is that, in weightlessness, airflow is your friend and keeps everything moving in the right direction. Once you have suction going into the hose, it's simply a case of maintaining good aim or, as I tell my two young boys at home, don't be a secret splasher!

For No. 2 there is a rather petite loo seat secured on top of a solid-waste container. This container has a small circular opening, around which is stretched a rubberised bag with an elasticated opening. Hundreds of tiny perforations in the bag allow air to flow through it, but not solid waste. The same switch on the urine hose activates airflow through the solid-waste container. On successful completion of business, astronauts drop the (self-sealing) rubberised bag with its contents into the container and leave a fresh bag in place for the next crew member. The solid-waste container is changed about every 10 to 15 days, although one ISS commander proudly told me that if you put on a sterile glove and pack down the contents, you can stretch this out to 20 days. I'm not sure the space programme fully appreciates the levels we go to, to conserve resources!

Air that is sucked through the loo is then dried, filtered and deodorised, prior to being returned to the living quarters. There are two toilets in the space station – one in the *Zvezda* module of the Russian segment and one in Node 3 (*Tranquility*) of the US segment. Despite being a fairly uncomplicated procedure, use of the space-station loo has not been without incident. One astronaut (who shall remain nameless) told me that following a call of nature one day, he turned around to dispose of the rubberised bag into the solid-waste container, only to find the bag completely empty. Feeling quite certain that it was supposed to contain a decent-sized portion of metabolic waste, there followed a fervent search for the liberated object. As with most missing items in space, it had vanished into thin air until two weeks later when, during routine maintenance on the loo, another crew member made a remarkable discovery of a small, hard and dried-up foreign body wedged into a small gap near the return air filter! This actually leads me nicely into the next question . . .

Q *What happens to waste from the space station?*

A Waste from the space station is loaded onto one of the resupply spacecraft that will undock at the end of its mission and burn up in Earth's atmosphere. A large amount of this waste is lightweight packing material. Since all items that are flown to space must endure the rigours of a rocket launch, plenty of foam or bubble wrap is used to protect the more delicate items.

We also throw away our used clothes, empty food packaging and solid-waste containers from the lavatory (think about that, next time you wish upon a shooting star!). However, we don't get rid of urine as it contains too much precious water. Instead, urine is recycled back into drinking water, and only the concentrated waste product (called 'brine' – nice!) is collected and eventually suffers the same fiery death as all our other rubbish.

Q *How does the space station get water and oxygen?*

A About 70–80 per cent of water on the ISS is produced from recycling (yes, along with urine, we also drink our recycled sweat and exhaled moisture!). This level of recycling is quite impressive and owes its success to a urine-and water-processing assembly, which filters impurities, removes all contaminants and produces drinkable water that is cleaner than what most of us drink on Earth. But as good as this system is, future space exploration beyond the relative ease of an Earth resupply is going to demand closer to 100 per cent recycling. The space station currently receives additional water supplies from visiting cargo spacecraft. There is sufficient water on the ISS for a daily intake of 3–4 litres per crew member, which is plenty for drinking, hygiene and the rehydration of food products.

Oxygen is generated by using water and a process called electrolysis, which separates liquid water into hydrogen gas and oxygen gas. Rather than simply wasting the hydrogen gas, there is an extremely clever device called 'Sabatier', which uses a catalyst that reacts with this hydrogen and carbon dioxide (which we breathe out) to make water again, leaving just methane as a waste product. There are also high-pressure tanks containing oxygen and nitrogen both inside and outside the space station. These are used if we need to replenish the ISS atmosphere, or to recharge our spacesuits following a spacewalk (we breathe 100 per cent oxygen inside our spacesuits).

Q *How long does it take to get used to floating in weightlessness?*

A The good news is that it doesn't take too long to become reasonably proficient at moving around in microgravity, although it did surprise me how clumsy I felt when I first arrived on the space station. There's a real skill involved in controlling your body in weightlessness (particularly your legs) and, on first entering the space station, I felt somewhat uncoordinated and aware that my legs were bumping into things. However, after about a week most astronauts will have mastered the basics of weightlessness (although to execute a perfect back-flip may take a while longer!). In that respect, 'floating' is much like any sport and, with time and practice, you can get really good at it.

The Russian segment is a good place to spend your first hour or so in space as the modules are of a smaller diameter, so you are never too far from a handrail to steady yourself and regain control. When you first float into the US segment the modules seem huge in comparison, and the walls very far apart. This is because the US modules were launched in the cargo bay of the Space Shuttle, which has a larger diameter than the Proton rocket that carried the Russian modules to space.

During one early foray into the Japanese *Kibo* laboratory, I accidentally let go and found myself floating in the middle of this large module, unable to reach a handrail. After several moments of unceremonious flailing of arms and legs, I was able to 'swim' my way to a handrail and avoid the embarrassment of having to call out for assistance from a crew member. Of course I'm sure my rookie struggle was caught on the ever-watchful cameras and probably caused much hilarity in Mission Control.

Q *What's the best bit about floating?*

A It's very relaxing and a wonderfully liberating sensation, as you don't have to work against Earth's gravity. Your muscles will naturally find their most relaxed position in weightlessness. If you do nothing, your body adopts a sort of bent posture, somewhere between sitting and standing, with slightly hunched shoulders, which tend to float up, along with your arms.

Another fun part of floating in microgravity is that you see everything from a new perspective – there is no upside down in space! You can use all of the available volume inside the space station for living and working, and it's as easy to flip around and work on the ceiling or walls as it is on the floor.

Also, you can store large or 'heavy' objects on the walls and overhead panels with just a weak bungee cord or some Velcro, without having to worry about it falling down. Floating gives you a *lot* more room in which to live and work. Imagine how much extra space you would have, if you could use all of the available volume in the room you are in right now.

Probably one of the best things about floating, though, is being able to push off with a hand or a foot and propel yourself from one end of a module to the other and, depending on how cocky you feel, maybe throw in a somersault or two, just for kicks!

Q *Why does the ISS use Greenwich Mean Time (GMT) in space?*

A In order to have a daily routine on board, it is vital that the ISS has a time zone. The space station runs to UTC (Universal Time Coordinated), which is equivalent to Greenwich Mean Time. When deciding the space station's time zone, Greenwich Mean Time was a compromise between the major participating nations of the ISS programme (US, Russia, Europe, Japan and Canada). By using GMT, most of these countries can overlap a reasonable part of their normal working day with some of the crew's duty hours on the ISS. Of course this time zone is great for the European Space Agency, but the Japanese Space Agency is probably the most disadvantaged – its Mission Control Centre is located in Tsukuba, just outside Tokyo, which is nine hours ahead of GMT.

In the past, the ISS time zone would sometimes shift to Florida time (Eastern Standard Time or GMT-5), when a Space Shuttle was launching from Kennedy Space Center, in order to support the Shuttle mission objectives. However, with the retirement of the Shuttle in 2011, the time zone on the ISS seldom changes, and for the duration of my mission it remained on GMT.

Q *What was your daily routine when there are 16 sunrises/sunsets every day?*

A Our daily routine was that of a normal 12-hour workday (7 a.m. to 7 p.m.), with no allowance made for the fact that we actually saw 16 sunrises and sunsets each day. As it happened, the rapidly changing day/night cycles were not difficult to get used to. Despite the fact that no two days on board the ISS were ever the same, with all the science experiments, maintenance activities and other tasks providing constant variety to the schedule, it was fairly easy to establish a stable routine.

I usually woke around 6 a.m., which gave me an hour for personal hygiene and breakfast prior to the morning brief. During this time I would also check to see if there had been any changes to the schedule. Where possible, I would prepare my activities the night before. However, the ISS is a very busy and dynamic environment to work in, and occasionally Mission Control would have to implement last-minute alterations to our schedule during crew sleep. This was also a good time to check which areas of interest to photograph that day. Occasionally science-based photos would be requested by the Earth observation programme. This might entail monitoring volcanic activity, documenting glacial retreat, asteroid impact craters, coastal regions or river deltas. Other targets were noted purely for personal interest – for example, an orbit sweeping across the Himalayas, clear weather over Europe (rare during the winter months!) or a direct overhead pass of the Pyramids would always be highly valued. With 16 orbits a day, there was never a shortage of places to photograph. I would set alarms for each of these targets, in the hope that I might catch one or two of them.

Work officially started at 7 a.m. with a quick morning brief, called the Daily Planning Conference or DPC. This was a round-robin-style check-in with all Mission Control Centres around the world, starting with Houston, then on to Huntsville (Alabama), Munich, Tsukuba (Japan) and finally Moscow. The DPC would last around 15 minutes, and then it was time to get on with the day's activities. Our tasks consisted mainly of facilitating scientific experiments. The crew's work schedule and detailed instructions for each task were displayed on several computers located around the space station, in addition to the fairly recent addition of personal iPads connected to the ISS Wi-Fi. Astronauts sometimes comment about being chased by a 'big red line', in reference to a red line displaying the current time that marches across the screen, leaving you in no doubt as to whether you are running ahead or behind schedule. Some days would involve executing one or two longer or more complex experiments, and on other days you might be involved in 10 or 20 smaller activities.

In addition to experiments, the crew would also conduct maintenance tasks, education/outreach activities and public-relations duties. Furthermore, whenever a cargo spacecraft visited the station, many hours would be spent either unpacking or packing the vehicle. This would be in addition to some refresher training ahead of time, for that all-important 'capture' using the space station's robotic arm. If a spacewalk was planned, it usually became the crew's primary focus a few days beforehand to ensure that all the equipment was ready and in good condition, and that the crew were fully prepared to execute what is arguably the highest-risk activity of any space mission.

There was normally an hour planned for lunch, which was always a good buffer in case some of the morning tasks overran. During the afternoon it would be more of the same, although my preference was to exercise late in the day, and therefore often from about 5 p.m. onwards I would be doing two hours of cardiovascular and strength training. The working day would finish at 7 p.m. with another round-robin call to all Mission Control Centres. Following this, the crew would usually eat a quick evening meal and prepare for the next day's activities. This usually left an hour or two at night to catch up on emails, make a few phone calls to friends and family or take some photographs, before getting to sleep around 11 p.m.

Q *How did going up into space affect your sense of time?*

A The human body has a natural perception of time, called the circadian rhythm. Our circadian rhythm not only dictates when we feel tired and alert, but many bodily functions are also affected in more subtle ways – for example, our body temperature, levels of concentration, cognitive ability and digestive system.

The most common way to mess up the body's circadian rhythm is to experience a large and rapid shift in time zone – more commonly known as 'jet lag'. Prior to launching to space, each crew will spend about four weeks in Moscow completing final Soyuz exams, followed by two weeks in quarantine at the launch site in Baikonur, Kazakhstan. However, during quarantine we remain on Moscow time, as it is easier to shift from Moscow to the space station's GMT (-3 hours) than it is from Baikonur to GMT (-6 hours). On arrival at the ISS, astronauts make an immediate shift to GMT, in addition to overcoming the effects of a very long launch day. Yuri, Tim and I had been awake for about 24

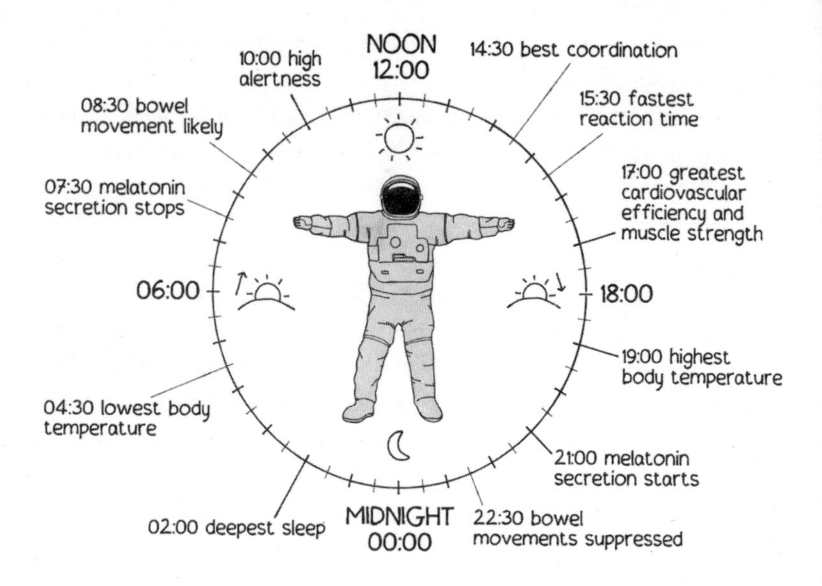

hours on launch day, so there was definitely an element of jet lag to overcome in the first couple of days in space.

Furthermore, our circadian rhythm is also very sensitive to light, in order to regulate our perception of time. One of the main challenges in adapting a good circadian rhythm in space is to get used to the frequent day/night cycle, caused by orbiting Earth 16 times a day. At first, this feels quite strange. It's certainly bizarre when you grab an 11 o'clock coffee break but it's pitch-black outside and you're over China, for instance, and then you're brushing your teeth at night and it could be bright daylight over Europe.

In fact I discovered the hard way that the worst thing to do, prior to bedtime, is to look out of the window during daylight. The massive influx of ultraviolet light that you get from the Sun stops the body producing melatonin (the hormone that makes you feel sleepy). This wreaks havoc with your circadian rhythm and stops you sleeping for hours. After making this mistake once, I was always very careful to ensure that I only looked out the window if we were in darkness prior to heading to bed.

However, the ISS is a busy place and it runs to a strict routine. This is good for promoting a strong circadian rhythm. I have also found that eating meals at regular times, and doing physical exercise at the same time each day, can help the body to adjust rapidly to a new time zone. After about two weeks in space my sense of time there was very good – I was sleeping well and was unaffected by the 16-orbit day/night cycle.

Interestingly, as our mission was coming to an end, the ISS was about to be fitted with new light-emitting diode (LED) lighting. The existing lights could be dimmed if too bright, but they were fixed-frequency 'white lights'. The new LED lights have the ability to shift frequency, adopting a white/blue light (shorter wavelength) for optimal performance during the working day, but shifting to a red light (longer wavelength) in the evenings prior to sleep. I'm sure this will prove to be a welcome modification among the crew.

Q *What's it like to sleep in space, and where do astronauts sleep?*

A Astronauts on board the ISS each have a designated crew quarter, which is about the size of a small shower cubicle. There are four crew quarters in Node 2 (*Harmony*) of the US segment and two crew quarters

in the Russian segment. I slept in the Node 2 'deck' crew quarter – the others being port, starboard and overhead. Of course in space there is no up or down, and so the crew quarters can be in any orientation. When it comes to the best technique for sleeping in space, opinions vary between astronauts. I liked to clip my sleeping bag very loosely to hooks in the wall, then zip myself into the bag and just let myself float. Our sleeping bags are quite close-fitting, which is good because you don't want to move around inside them too much. You have the option of putting your arms through slots in the sleeping bag and 'wearing' it like a long sleeveless robe. However, that raises an even bigger dilemma of what to do with two unrestrained arms that float around all night, with the potential to bump into things and wake you up.

Some astronauts prefer to be strapped more securely to a wall in their crew quarter, and others elect to be completely detached and float around all night – at the risk of bouncing off a wall in the middle of the night! The biggest hurdle to overcome initially is falling asleep. Astronauts spend all day floating and, when it's time to sleep in space, your body does not receive the usual triggers of lying down and putting your head on a pillow to help you fall asleep. All you can do is turn off the lights and carry on floating . . . whilst waiting for the body to give in to slumber. Astronauts have even been known to strap a makeshift pillow to their head to help give the body that extra trigger to encourage it to sleep.

It took me a couple of weeks until I could fall asleep easily, but after that sleeping was not an issue. I usually got between six and seven hours' sleep a night, but I'm not convinced it was of such good quality as I get on Earth. It's sometimes hard to get your arms into a comfortable position, as they want to float up and out in front of your body. I would fold them across my chest and use the tightness of the sleeping bag to help keep them in one place. I sometimes used earplugs to aid sleep and deaden the constant hum from the ventilation fans, which are needed to keep air circulating and prevent dangerous pockets of carbon dioxide from building up inside our crew quarters.

It's ironic that weightlessness should be such a wonderfully liberating

My crew quarters, located in the deck of the Harmony module.

feeling and yet it deprives us of gravity's one consolation prize – collapsing into bed at the end of a long day and feeling the weight of your head on a soft pillow. Now that is something I really missed whilst in space!

Q *Do the astronauts all sleep at the same time?*

A This is an interesting question because many people might assume that someone has to be awake at all times on the space station. In the Army we would call this 'stagging on', with everyone taking a turn at a couple of hours' watch-duty during the night. However, on the space station the crew all sleep at pretty much the same time, anywhere between around 10 p.m. and 6 a.m. The ISS commander has an emergency alarm in their crew quarter, and in case of an emergency they are responsible for waking the rest of the crew. Of course for the majority of its orbit the ISS is in communication with multiple Mission Control Centres around the world, and so in reality there are many people on Earth closely monitoring the space station whilst the crew sleep.

Q *Did you dream differently in space, or dream of anything in particular?*

A I don't dream that much. More accurately, I suppose I should say I don't often remember my dreams, since some experts say that we all dream at least four to six times per night. I've heard friends and family sometimes recount their dreams in the most amazingly vivid detail and have often wished that I could do the same. The few dreams that I do remember whilst in space were of places back on Earth, and I would be walking around in normal gravity. There was one exception. Towards the end of my six-month mission I had a dream where I was searching for a book in a library that had really tall bookshelves, stretching high up to the ceiling. I remember being frustrated at not being able to reach the highest shelves, wondering why there were no ladders around. It suddenly occurred to me that this wasn't a problem at all – after all, I could just float up and search the shelves, couldn't I? Suddenly I was weightless in my quest to find this elusive book, and it seemed perfectly normal to be floating around back on Earth. I never did find that book, though.

Q *Which has been your favourite experiment, and why? – Adam, from Castell Alun High School in Flintshire*

A That's a tough question for me to answer because during my six-month mission we completed well over 250 experiments and there were many highlights. Generally speaking, I really enjoyed conducting life-science experiments. This is because, as an astronaut, you are the subject of these investigations. Life-science experiments often involved performing medical procedures in space, which were new and interesting to me. I became pretty good at self-phlebotomy (blood-draws), ultrasound (eyes, heart, arteries, veins, muscles), ophthalmoscopy and optical coherence tomography (eye imaging), tonometry (measuring the fluid pressure inside the eye), corneometry (skin hydration) . . . the list goes on. Of course the downside of participating in so many life-science experiments was the requirement for frequent bodily 'donations' of urine, saliva, faeces, breath and blood. These studies are not only conducted in space, but often commence well before launch and can last for two years or more after landing, in order to fully track the impact of spaceflight on our bodies. One experiment also required painful muscle biopsies pre- and post-mission.

Then there were experiments that drew on my prior experience as a test pilot in evaluating equipment, such as controlling a 'martian' rover on Earth from the ISS. My task was to use the rover to explore a dark cave, which had been constructed at the Airbus 'Mars Yard' in Stevenage. I had to identify various rocks and features whilst evaluating the 'human-machine' interface and a novel communications link. This will help pave the way for future astronauts en route to Mars to be able to control vehicles on the Red Planet in preparation for their arrival, and for astronauts to control rovers on the surface of the Moon whilst in lunar orbit.

If I had to narrow down my choice to one favourite experiment, I would pick ESA's 'Airway Monitoring' investigation. This was a complex experiment executed over a couple of days. It involved using the space station's airlock as a hypobaric chamber – a first for ISS science. Astronauts are exposed to higher levels of dust in space, since it doesn't settle to the ground. For future space exploration this situation could be much worse. For example, exposure to dust storms on Mars would be very harmful, as would inhalation of the fine regolith covering the surface of the Moon (with no weather on the Moon to smooth these minute rocks, their sharp, jagged edges would do untold damage to your lungs if inhaled). In space, fine dust can cause irritation to the eyes, inflamed lungs and asthma, just as it does on Earth. Every time we breathe out, we expel small amounts of the gas nitric oxide (NO), which is produced by our body to regulate blood vessels and act as an antibacterial agent. Doctors can use this expelled NO as an indicator of when our airway has become inflamed. The Airway Monitoring experiment analysed the quantity of NO expelled under a variety of conditions, including breathing at reduced pressure in the airlock. This ground-breaking research into lung physiology will be used to benefit not only future space explorers, but also millions of asthma-sufferers on Earth. That gets my vote as a first-class experiment, and it's also a great way to introduce the next question . . .

Q *What are the benefits of research done in space?*

A When Yuri Gagarin launched to space on 12 April 1961, there was real concern amongst flight surgeons that the human body would not tolerate weightlessness. The possibilities for catastrophe were numerous, with heart, lung and brain malfunction high on the list. Since then, humans have not only tolerated weightlessness, but have managed to adapt and excel in this new environment for extended periods of time. Along the way, a plethora of information has been learnt – not only about the human body, but relating to almost every field of scientific research. And it's not only government-funded research that is being conducted. More and more commercial companies are realising the benefits of space-based research, as the ISS grows as a platform for innovation within industry and the private sector. Rather than making generic statements about how ISS research has benefited our daily lives, I hope you'll find it interesting to read a few of the many concrete examples where space-based research has been a game-changer for people on Earth.

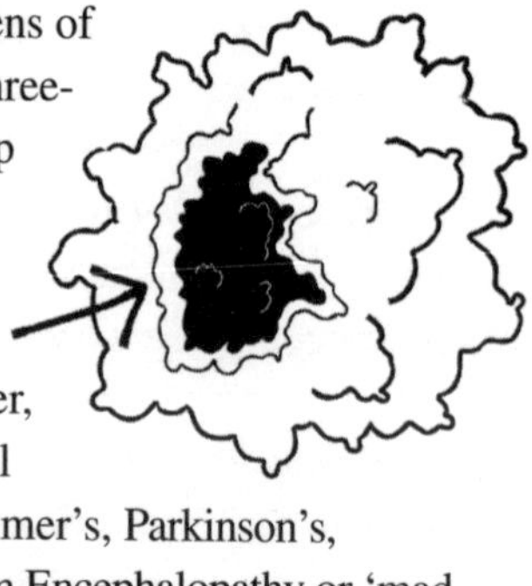

Protein structures: Our body contains tens of thousands of different proteins. These three-dimensional complex structures make up nearly 17 per cent of our total body weight. They form the structure of our bodies and play an important role in the processes that keep us alive. However, proteins can 'misfold' and act in harmful ways, leading to diseases such as Alzheimer's, Parkinson's, Huntington's and even Bovine Spongiform Encephalopathy or 'mad cow' disease. Most drugs used to treat these diseases release small molecules designed to fit into the folds of the disease-causing proteins and inhibit their function. But, in order to work effectively, the drug-molecule needs to fit the misfolded protein accurately, like two pieces of a three-dimensional jigsaw puzzle. This requires a detailed knowledge of the protein's structure. Without this knowledge, higher doses of an inferior drug are often used instead, which can in turn produce harmful side-effects. The protein's structure can be investigated by developing a protein crystal and using a technique called X-ray crystallography.

The key to success lies in harvesting high-quality protein crystals. Researchers have found that it is easier to grow high-quality crystals in weightlessness, where they form more slowly, without the effects of gravity and convection that distort or destroy the crystals' delicate structure. Space-grown crystals have proven to be larger and more perfect than any obtained on Earth, and have already enabled valuable insights into treatment for Duchenne muscular dystrophy. New targets, including Hepatitis C, Huntington's disease, some cancers and cystic fibrosis are currently being tested in space. This barely scratches the surface of the potential for this research. With as many as ten billion proteins in nature, with every structure being different and holding important information related to our health and the global environment, this is one of the most exciting areas of ISS research.

Vaccine development: The space environment causes a multitude of changes in microbial cells, such as alterations of microbial growth rates, antibiotic resistance, microbial invasion of host tissue and even genetic changes within the microbe. One change due to microgravity, which is of particular interest in terms of studying infectious diseases, is virulence (the ability

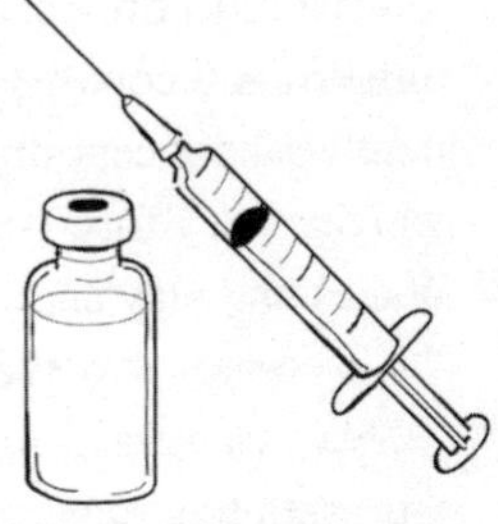

of a microbe to cause disease). This has been shown to increase in microgravity. Scientists have been able to use microgravity to identify those viral strains that are least potent, which are then selected as candidates for use in vaccine development on Earth.

Salmonella is one of the most common forms of food poisoning, with *Salmonella* diarrhoea remaining one of the top three causes of infant mortality worldwide. The space-based research of commercial corporation Astrogenetix has resulted in the discovery of a potential candidate vaccine for *Salmonella*, now in the planning stages for review and commercial development. Follow-on experiments on the ISS have examined the virulence of methicillin-resistant *Staphylococcus aureus*, more commonly known as MRSA. More recently still, samples have been flown to the space station that seek to improve on existing vaccines against *Streptococcus* pneumonia – a bacteria responsible for more than ten million deaths annually, and which causes diseases such as pneumonia, meningitis and bacteraemia.

These results represent just a fraction of the possibilities of future microgravity vaccine development. Scientists participating in these studies plan to fly a series of future experiments to the space station that will help to accelerate the progress of several different life-saving vaccines.

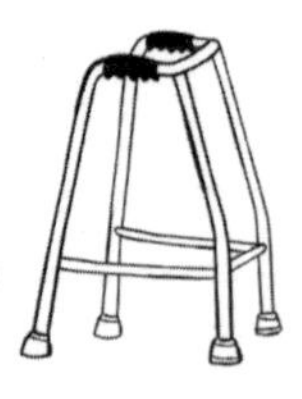

Ageing process: Rapid changes that occur to the human body as it adapts to microgravity provide a unique model to study the ageing process (without the actual complexity of age!). This includes research into bone loss, cardiovascular degeneration and changes to skin, balance and the immune system. Studies have already led to the development of a new drug for osteoporosis (Amgen's Prolia™), with many more ISS experiments currently targeting this area of research. The world's older population is growing at an unprecedented rate, with 8.5 per cent of people worldwide now aged 65 and over. In the US alone, the number of people in this age group is set to double over the next three decades. However, living longer does not necessarily mean living healthier. ISS research is helping to prepare for the public-health challenges posed by this increase in our older population, to reduce the financial burden of age-related disease and improve quality of life for the elderly.

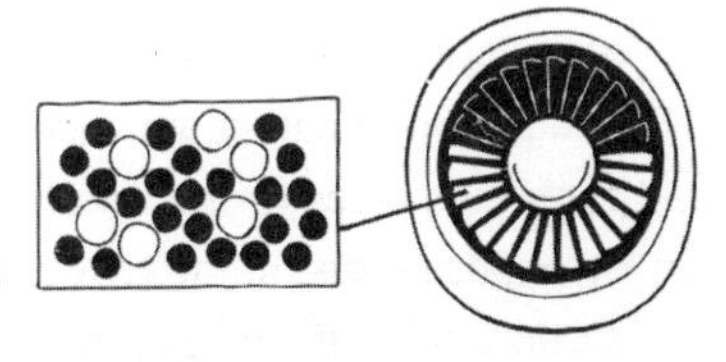

Metal alloys: Casting has been around for a while – in fact the oldest-surviving cast is a copper frog from 3200 BC. However, whilst forcing molten metals to mix with other elements in a cast may not be a revolutionary idea, it is still producing novel results and is at the forefront of cutting-edge research. On Earth, as a newly created metal alloy cools and crystallises, the microstructure suffers from convection and sedimentation, due to gravity. Understanding the physical principles that govern this solidification process is crucial to producing high-quality materials such as solar cells, thermoelectrics and metal alloys. In space, the absence of convection and sedimentation allows scientists to control solidification, in order to better understand the process and improve ground-based metal casting, leading to new, stronger, lightweight materials. The European 'Electro-Magnetic Levitator' or EML payload on board the ISS is one of the leading experiments in this field. An ESA-led project called IMPRESS combined 43 research groups from academia, industry and the science community, to develop turbine blades manufactured from titanium aluminides. These crystalline alloys offering unique qualities, such as a high melting point, high strength and low density, are ideally suited for modern power stations and aero-engines. Using titanium aluminide would result in a 50 per cent weight reduction of turbine components, leading to higher efficiency, reduced fuel consumption and lower exhaust emissions.

Cold plasma: Plasma is one of the four fundamental states of matter – the others being solid, liquid and gas. A plasma is an electrically charged gas, a bit like lightning, which rarely occurs on Earth. In contrast, 99 per cent of the visible matter in space is in the plasma state. When dust particles or other microparticles are also contained in the ionised gas, they become highly charged and a 'complex plasma' is formed. The ISS provides ideal conditions for the investigation of complex plasmas because, in microgravity, the dust particles can spread freely in space and form ordered three-dimensional plasma crystal structures. This has provided a whole new insight into physics.

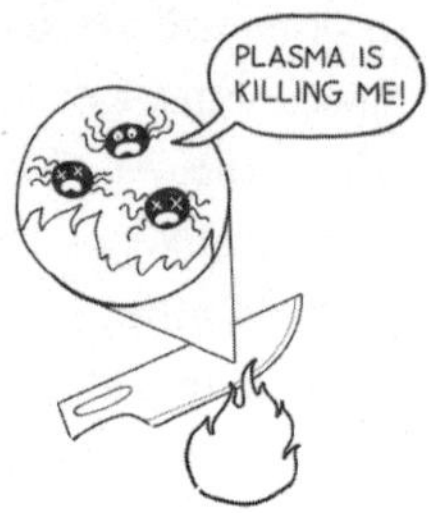

Plasma can permeate many materials, spreading evenly and quickly. It can disinfect surfaces and has been proven to neutralise drug-resistant bacteria like MRSA within seconds. Plasmas can also disinfect chronic wounds, and help wounds to heal faster. Other research has shown that (along with chemotherapy) plasma treatment efficiently fights cancer, boosting tumour inhibition by 500 per cent, compared with chemotherapy alone.

Following a hugely successful series of European-Russian experiments on the ISS and a subsequent technology-transfer programme, this knowledge of complex plasma has developed into practical 'cold plasma' applications on Earth. Since 2013, the Terraplasma company has successfully applied cold-plasma technology to treat many medical and hygiene problems, in addition to water treatment.

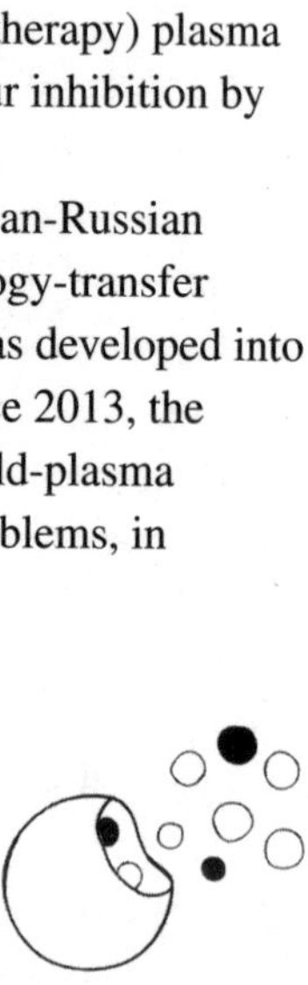

Microencapsulation: Imagine filling a tiny, biodegradable micro-balloon (about the size of a red blood cell) with various drug solutions that can be injected into the bloodstream to fight disease; inhaled to treat bacterial infections of the lungs; or delivered directly to the site of cancerous tumours. This is a process called microencapsulation and it has been developed using space-based research, with remarkable results. Tests have shown that a few doses of microcapsules injected directly into human prostate tumours have inhibited growth by up to 51 per cent within three weeks. Other studies showed that just two doses of microcapsules injected into lung tumours produced a visible size reduction in 43 per cent of the tumours. After 26 days, lung-tumour growth was inhibited by 82 per cent, and 28 per cent of the tumours had completely disappeared. These positive results with such low doses have clear advantages over traditional cancer treatments, such as chemotherapy.

The ISS was vital to the development of these capsules because microgravity enables dissimilar liquids (such as oil and water) to disperse evenly throughout a globule. This enables the pharmaceutical and its outer membrane to form spontaneously, producing extremely high-quality microcapsules.

Did you know?

- If you have ever tried to make an oil-and-vinegar salad dressing, you'll know that these two liquids are *immiscible*, meaning that they don't mix. The oil floats on top of the vinegar because it's less dense. You have to shake the bottle in order to get an even amount of oil and vinegar on your salad. After a short time the oil and vinegar will separate again, because of the effect of Earth's gravity on the differing densities of the liquid. This is not the case in space. On the ISS you can enjoy your oil-and-vinegar dressing without having to shake the bottle! The two liquids are still not truly mixed, but without the effect of gravity, small globules of oil and vinegar disperse fairly evenly within one larger globule. This is the basic principle behind microencapsulation in microgravity.

The resounding success of the space-produced microcapsules led to the invention and NASA patent of a Pulse Flow Microencapsulation System, or PFMS, which is an Earth-based system that can now replicate the quality of the microcapsules created in space.

And it is not just cancer patients who could benefit from this method of treatment. Microcapsules could be used to treat several diseases – for example, to eliminate daily insulin injections, as diabetes sufferers could use implanted microcapsules as a method of controlled release over 12–14 days.

Q *Was there a favourite part of your day in space?*

A I always enjoyed wrapping up at the end of the day and then having the chance to take some photographs, look out of the window or call friends and family. The working day was very rewarding, but we were always trying to keep up with a tight timeline and, when we did get ahead, there was a list of other tasks waiting to be performed.

That's good, though: a busy schedule makes for a demanding and energising environment, and we are in space to work hard! I am amazed when people ask me if life on the ISS was ever dull . . . On the contrary, you relish the few quiet moments that you have to yourself. It may sound strange, but I would always enjoy brushing my teeth in the evening, because our hygiene area was close to the Cupola, so I would float next to this enormous window and enjoy the view for a few quiet moments. I loved that juxtaposition of doing the most mundane task whilst travelling at 25 times the speed of sound and watching an entire continent passing beneath me.

Q *Do you have time off? How do you spend your weekends?*

A During the week there was precious little spare time, which made the time pass incredibly quickly. The weekends had a more relaxed tempo, and we would usually manage a few hours to ourselves. Saturday mornings were spent cleaning the space station. A lot of dust collects in the return air filters of the ISS, and it takes a couple of hours for the crew to vacuum them clean. I suppose it's quite humorous to think of astronauts doing the hoovering in space, but hey – no one else is going to keep the place clean! We actually used an ordinary off-the-shelf AC vacuum cleaner, with one of the longest extension cords I have ever seen. I remember one of my favourite books, growing up, was *The Flying Hockey Stick* by Roger Bradfield. This boy flew around the countryside on a hockey stick powered by an electric fan, reeling out cable behind him as he went about his adventures. Hoovering the ISS was a similar endeavour – flying through the space station with a vacuum cleaner unreeling endless cable.

As well as being good housekeeping, it was also a great way to find any items that had been lost during the week, because the fans would slowly but inexorably drag everything towards the return air filters. I remember Tim Kopra presenting me with a '100 days in space' patch. I let go of it momentarily to grab a camera and, sure enough, when I turned back, it had disappeared (you'd think that after 100 days I might have learnt that lesson already). The patch turned up two weeks later, stuck to a return air filter in an adjacent module.

In addition to vacuuming, we would also use disinfectant wipes on surface panels, handrails and anything we would come into contact with, in order to keep microbial growth to a minimum and reduce the risk of infection. Each crew does the cleaning slightly differently, but we decided to split the US segment into three large modules for each person and rotate every two weeks – that way, one person was not stuck with cleaning the loo for six months! After housekeeping, I would dedicate my Saturday afternoons to education outreach projects. Sometimes this would involve recording messages or managing student science experiments, such as running students' code on Raspberry Pi's space-hardened 'Astro Pi' computer. At other times it involved talking to school children over ham radio or hosting events from space, such as the 'Cosmic Classroom', which was watched by half a million students.

Ideally, Sundays would be free time, and each crew member had the opportunity to have a short video-conference with family members.

This was incredibly important for crew morale, to be able to maintain connectivity with your loved ones, despite being so remote and detached from planet Earth. Often there would be tasks that still needed doing at the weekend and we had to exercise, of course, but generally there was some free time for catching up on photography or calling friends and family.

Q *What's the grossest thing about living in space?*

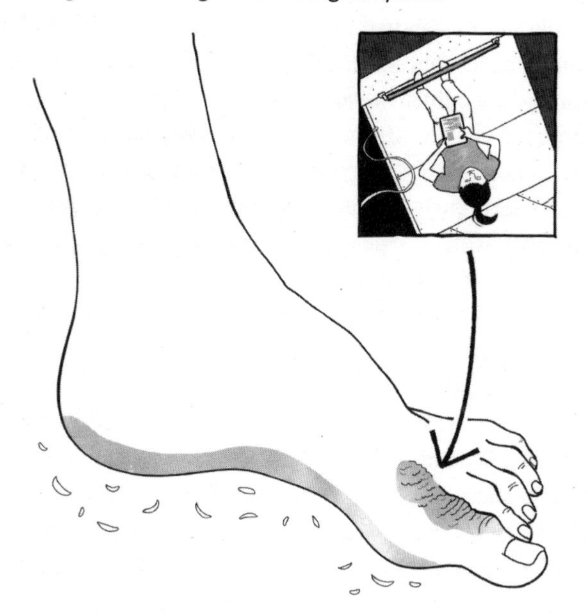

A Ha, what a great question! By far the grossest thing about living in space is watching the soles of your feet disintegrate during the first couple of months in space. We hardly use the soles of our feet on the space station, and there is seldom any weight on them (except when we exercise). Because of this, they become very smooth and soft, like a newborn baby's. Six months in space is akin to having the best pedicure you could imagine.

The gross part is that all the dead, hard skin that builds up on the soles of your feet starts coming off. After living in space for a few weeks you have to take your socks off very carefully, otherwise there will be a shower of dead skin-flakes ejected into the cabin. As nothing sinks to the floor in microgravity, this skin would just hang around until the airflow gradually pulled it towards one of the return air filters. Meanwhile you would rapidly become the least popular member of the crew!

Equally gross is the fact that we develop 'lizard feet' on the tops of our toes. We are constantly hooking our feet underneath metal handrails, straps and bungees and using this force to hold us down and stabilise our body position whilst we work. All of this abrasion causes the skin on the tops of our toes to become very rough and scaly. In fact the European Space Agency has even experimented with specially designed socks, in an effort to prevent this. The socks have a soft rubber coating over the top of the toes and do help, to some degree.

Q *Did you have any personal reading material, and what would be your choice of book to read in space?*

A Yes, astronauts can read e-books in space, if they wish, or (more commonly) listen to audio books. We have a crew support team on the ground that will, amongst many other things, send e-books, podcasts, news articles, music files and even TV programmes via the communications data link-up to the space station, if requested.

I didn't read much whilst in space, mainly because what limited free time I had at the weekends and evenings was spent taking photographs or calling friends and family. Whilst exercising, I would often catch up on the news or listen to podcasts (*The Infinite Monkey Cage* with Brian Cox and Robin Ince being one of my favourites, or the *Chris Evans Breakfast Show*). However, I did take an original hard copy of Yuri Gagarin's autobiography, *Road to the Stars*, with me. This book belongs to Helen Sharman and was signed by Gagarin himself, along with Helen's crew during their mission to the Mir space station in 1991. I can't think of a better choice of book to read whilst in space – it was a real honour to be able to borrow this book from Helen, and a memorable experience reading it up there.

Q *What surprised you most about the space station?*

A By the time you actually arrive on the ISS, you are so well trained and have spent so many hours studying every possible aspect of the space station that there are few, if any, surprises. That's not to say the ISS doesn't have its fair share of quirks and flaws – it's just that astronauts usually discover them as they progress through training. One of the most surprising facts I learnt in the early days of training was that the Russian segment and the US segment operate on different electrical voltages. Both segments receive 100 per cent of their electricity either directly from the space station's solar panels or (during darkness) from batteries charged by the solar panels. However, the Russian segment converts this solar power to 28 V DC, whereas the US segment converts it to 124 V DC.

This is not such a startling fact in itself, but the knock-on effect is that the Russian fire extinguishers (based on a water/foam mix) are prohibited from use in the US segment, due to the risk of electric shock. Instead, the US segment uses carbon-dioxide fire extinguishers, but these are prohibited from use in the Russian segment because, in the event of a discharge, the Russian life-support systems are not designed to cope with scrubbing such large quantities of carbon dioxide (by 'scrubbing' I mean removing carbon dioxide from the air so that it can be breathed again). And it's not only the fire extinguishers that are different.

One of the major success stories of the ISS is the fact that 15 nations collaborated to construct and operate the most complex engineering project in history. However, with so many different nations and companies involved in its construction, there are a myriad of 'non-standard' elements between different segments and modules. These range from the trivial (switches, fasteners, nomenclature, etc.) to the more important (emergency equipment, communications systems, life support, etc.). As a test pilot, I had become very familiar with evaluating aircraft and identifying those areas that place an increased burden on the crew by forcing them to 'work around' or compensate for a design flaw. Let's just say that I'm glad I didn't have to write the 'human factors' report for the ISS! As an example, one quirk of the communications system is that, in the event of a caution, warning or emergency occurring, all voice communication from the Russian segment to the US segment is disabled whilst the audio alarm is going off. Only by silencing the audio tone can you re-establish that communication link and hear what your Russian colleagues have to say . . . if they haven't already left in the Soyuz without you!

Q *Can you drink a cup of tea in space? – Katie Loughnane*

A This is essential information for any British astronaut, and you'll be pleased to know that *yes*, we can enjoy a cuppa in space! Actually, NASA lets us choose three hot drinks each day, which are tailored to our liking. I decided on two teas and one coffee. The good news was that NASA was able to certify my favourite brew fit for space travel. I'm a fan of Yorkshire Red (good builder's tea) and so, having passed the rigorous microbiological testing, my teabags were vacuum-sealed in a foil pouch (which serves as our drinking vessel), along with powdered creamer ('Heresy,' I hear you cry, but that was the only option) and a bit of sugar. To enjoy our hot drinks, all we needed to do was add hot water to the foil pouch from the Potable Water Dispenser (PWD) and sip it through a straw. We cannot drink from an ordinary cup or a mug because unfortunately in microgravity the hot liquid would just float away and cause a terrible mess!

When I was told about this tea-drinking process prior to flight, one immediate problem that sprang to mind was: how can you control the strength of the brew? It would be too weak to drink immediately, but by the end of the drink you would be sucking on a tea bag (ugh!). So, with a bit of tweaking, I was able to modify a drinking straw into a transfer tube that would allow me to let the tea stew to perfection, then transfer it to an empty foil pouch, so that I could enjoy drinking it at my leisure. All things considered, it tasted pretty good, for a brew made with creamer and yesterday's recycled urine!

NASA astronaut Don Pettit went one step further in 2008 when he designed a low-gravity cup for use during his mission. Don is quite simply a genius. He used mathematical modelling to determine the precise shape of a cup that would enable the fluid to be contained in microgravity, without spilling the contents all over the space station. The cup has a corner with a sharp angle, which, due to the surface tension of the fluid, acts like a wick and directs it towards the astronaut's

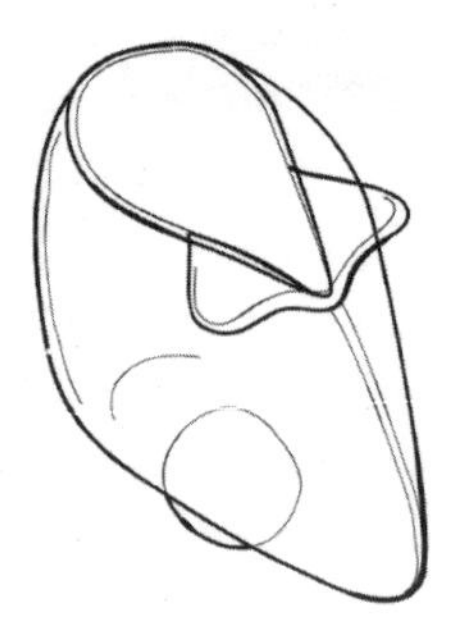

Don Pettit's cup

mouth. When you go to drink your coffee or tea, a capillary connection is formed and – hey presto – you can sip your hot drink in space! Whilst I did try this a couple of times for fun, I was never comfortable leaving a cup of hot liquid stuck to the wall with Velcro, so usually I would opt for the much safer option of a foil pouch instead.

Q *Did you watch movies in space?*

A Time is so precious on board the International Space Station that watching movies is usually way down the list of priorities. However, as a crew we probably watched two or three films together, which was a great way of winding down and relaxing at the weekend. The most memorable of these occasions was watching *Star Wars: The Force Awakens*. NASA astronaut Scott Kelly had asked for the movie to be sent up ahead of our arrival in December 2015. Mission Control can send up large data files over the satellite communication link, and on special occasions this could include a newly released movie.

A few months earlier Scott had helped persuade NASA management that the astronauts could really do with a projector and a large white screen on the ISS. This would enable us to conduct briefings, video-conferences and training events without the need for several people to be crowded around a small computer screen. That part was true enough, and the projector and screen are now used on a daily basis; but of greater interest to the astronauts – they're also perfect for movie nights.

Most of our films, TV shows and documentaries were not sent up over the communications link, but instead were pre-stored on a 1TB hard drive on the ISS. I had to smile when I first browsed this library and came across *Alien* – someone in crew support obviously had a sense of humour! *Gravity* was there too, but having watched it not long before launch, I didn't need reminding of the dangers of spacewalking just prior to my own spacewalk.

What's strange about microgravity is that although you can float comfortably in many positions, the most natural seemed to be an almost sitting position, and so for movie nights we would all find somewhere to

'sit' and enjoy the entertainment. Watching *Star Wars* definitely ranked extremely high on the cool factor. Enjoying an intergalactic battle whilst on a spacecraft orbiting Earth . . . I was half expecting to see a TIE fighter when I went to close the Cupola window covers that night!

Q *How do you wash your clothes in space?*

A There is no washing machine on the space station and water is a very precious resource, so we just wear the same clothes for several days, before we exchange them for a new item and throw the old ones away. It is not as bad as it sounds. We live in a temperature-controlled environment, so clothes do not get as dirty as they might on Earth. Some of the items, like socks and our exercise gear, sometimes have antibacterial materials in them, too.

We change our clothes according to a set schedule. This ensures that we will have enough items to last us the six months. For example, we change underwear every two to three days, T-shirt and socks every week, and trousers or shorts every month. Then we have a few extra polo shirts for smarter occasions (such as recording video messages and other public-relations activities) and a couple of sweatshirts, as the space station can sometimes feel cool in the evenings.

The clothing that is worked the hardest is our exercise gear. We change it every week, but when you're working out for two hours a day, then you certainly welcome a fresh set, come the weekend.

Q *Does your heart beat the same on the ISS as it does on Earth?*

A Studies have found that astronauts' hearts usually beat slightly slower in space than on Earth. This is because the heart muscle is not having to work as hard as it does when pumping blood against the effects of gravity. Not only is the heart offloaded in weightlessness, but our slightly reduced blood volume shifts upwards in microgravity towards our chests, making it even easier for the heart to do its job. The problem is that, like any muscle, the heart will lose mass unless it is exercised properly. Investigations have shown that some astronauts' hearts became more spherical in shape, as they atrophied in space. Thankfully, these changes were temporary, and the astronauts' hearts reverted to their normal mass and elongated shape, soon after returning to Earth. By studying these changes, researchers can fine-tune the correct exercises required to stay healthy on long-duration missions, which will be essential for trips to the Moon and Mars. More importantly, learning about the heart has numerous health benefits for people here on Earth.

Q *How do you cut your hair and shave in space?*

A Cutting hair in space is actually remarkably easy. We use a set of hair clippers that have been modified to include a rubber-tube attachment to a vacuum cleaner. The vacuum cleaner collects all the hair shavings (provided you remember to switch it on first!) and prevents them floating around the space station. I cut my own hair every two weeks in space – not something I have tried to repeat back on Earth!

For shaving, we have the option of using either an electric razor or a regular razor. When using the electric razor, it's best to shave next to a return air filter, in order to catch any stray whiskers. For wet shaving, we simply use warm water and shaving foam. The surface-tension property of water means that it tends to stick to your skin anyway, so you can shave pretty normally. Of course there is no sink or running water to rinse a razor in, so instead astronauts just wipe the blade clean using a facecloth.

I used an electric razor during the week, for ease of use and efficiency of time, but enjoyed a wet shave at the weekends.

Q *What's the atmosphere on the ISS?*

A The space-station modules are pressurised to a standard 'one atmosphere' – that is to say, the same pressure that you feel when standing at sea level on Earth (101.3 kPa or 14.7 psi). This makes life very comfortable for the astronauts, and is in fact better than you will experience on most aeroplanes. As you climb higher through the atmosphere, the pressure outside reduces and pressurised aircraft or spacecraft require greater structural strength to withstand this pressure differential. Greater structural strength often means more weight and greater cost, which is one of the reasons aeroplanes typically pressurise their cabins to the equivalent air pressure at around 1,830–2,400 metres.

The space station's atmospheric composition is also fairly normal, at around 21 per cent oxygen and 78 per cent nitrogen. This is much safer than using an atmosphere of pure oxygen, which would create an enormous fire risk. This tragically occurred during a ground test of the Apollo 1 capsule, when an electrical fault in an oxygen-rich atmosphere caused a rapid and extremely intense fire, killing all three crew.

However, the main difference between the ISS atmosphere and Earth's atmosphere is the higher level of carbon dioxide – more than ten times higher, in fact. This is simply due to the maximum effectiveness of the carbon-dioxide scrubbers on the ISS. The life-support systems on board the space station are a valuable resource and, whilst the carbon-dioxide levels could be lowered for short periods of time, it would take a toll on the longevity of these systems. For that reason, the level of carbon dioxide is constantly balanced between crew comfort and the management of resources. Whilst these higher levels of carbon dioxide are still within a safe regime, they can cause occasional headaches and a lack of mental sharpness. Human physiology is also a factor here, with some crew members being more susceptible than others to symptoms of high carbon dioxide.

Q *What is your favourite button on the ISS, and what does it do?*

A I love this question about the button. In the Japanese laboratory there's an airlock that we use to pass experiments to and from space. This enables us not only to study the microgravity environment inside the space station, but also to learn about the thermal extremes, radiation environment and, of course, vacuum environment outside the space station. During my mission we also passed several small satellites through this airlock and launched them into orbit, using the space station's robotic arm.

There's something very cool about opening up a door that leads to space! My favourite button on the ISS was the 'Open Outer Hatch' switch. I never lost the fascination of peering through the small airlock window and watching the outer hatch slowly reveal the vast blackness of space that lay beyond it.

As you would expect, we have some pretty impressive buttons in the Soyuz spacecraft, too. The really important ones have spring-loaded metal covers over them. It's a bit like the 'Are you sure you want to press Delete?' function on your computer – you know something irreversible is about to happen, so you had better be very certain what you are doing before pressing a button. I'm a big fan of these covers, having inadvertently pressed a few 'wrong buttons' in my aviation career – I suppose that's the downside of having a test pilot's curiosity.

In fact these Soyuz buttons are so important that they are hard-wired to their respective mechanical components, so that they will work even if there is a major computer malfunction. My favourite Soyuz button was the one that separates the spacecraft into three parts, when we're getting ready to enter Earth's atmosphere. Under normal circumstances this occurs automatically, but the crew have a manual backup, in case things don't go according to plan. When separation occurs, it is like a heavy machine gun firing right next to your head, as dozens of pyrotechnic bolts fire, splitting the Soyuz into three modules. Only one of these modules has a heat shield and, if all goes well after separation, it should be the one that you're still sitting in.

Q *What was your favourite pastime in space?*

A Without a doubt, my favourite pastime in space was photography. This surprised me, as I was not a keen photographer prior to my mission, and my holiday snaps were rarely worth a second glance. But viewing our beautiful planet from space is a rare privilege and, since the space agencies invest so much time, effort and money on training astronauts, it was no surprise that we received a comprehensive package of photography training before our mission. By the time I launched to space I was reasonably competent at handling our camera of choice on board the ISS, the Nikon D4.

Of course it's one thing to know the technical aspects of a camera or the theory behind photography, but actually taking a good shot from space is another matter altogether, and this largely boiled down to 'on the job' learning. For this, I owe a huge debt of gratitude to fellow crewmates Scott Kelly, Tim Kopra and Jeff Williams, all experienced astronauts who were willing to share their expertise. Some aspects of photography were much easier in space; we were provided with high-quality cameras and lenses, pure, unfiltered white light from the Sun and the most stunning subject in the solar system – Earth! However, moving at ten times the speed of a bullet often made things tricky, with very little time to identify and photograph targets. Night photography also brought its own challenges in managing low light conditions and achieving a good focus.

Often, to capture those more elusive shots of volcanoes, pyramids, glaciers or cities required meticulous planning. This was not just a case of knowing when you would be passing overhead, but of thinking about lighting conditions, oblique angles, optimum space-station windows to use and, naturally, the weather conditions down below. Whilst this took time and patience to perfect, the rewards were spectacular. I never imagined I would gain such a huge sense of satisfaction from some of the pictures I took. One of my favourite photographs has to be a rare shot of Antarctica, which is so far south of the ISS orbit that it's extremely unusual to get a clear shot (see photograph 29).

Q *What kind of food do you eat in space?*

A We eat fairly normal food, like you might eat on Earth, but clearly there are special requirements that make some foods more suitable than others for space travel. First, food has to travel to space, and so it must have robust packaging that won't break or burst during launch. Also food has to have a long shelf-life to prevent perishing or spoilage – usually a minimum of 18–24 months after packaging. Naturally food must be nutritious and healthy, with a balanced supply of vitamins and minerals. Finally there are some foods that just don't work well in microgravity: no matter how much I might enjoy the occasional packet of crisps, it creates way too much mess trying to eat them in weightlessness, and so crumbly food was off the menu.

Most space food is packed in foil pouches, plastic packets or metal cans and there are more than 100 items on the 'space menu' to choose from, so there's plenty of variety. Some of it is dried or freeze-dried food, which we can rehydrate with hot water to make it edible. Many of our vegetables and soups are rehydrated. Other (irradiated) food comes in foil pouches, which we place in our electrical food-heater for about 20 minutes to warm up. These foil pouches often contain our meats and desserts. They are a bit like military rations or camping food and don't taste too bad, although they can be a bit bland, as the salt content for space food is often reduced. Studies have determined that the skin retains sodium in microgravity, which increases acidity in the body and can accelerate bone loss – not good news for astronauts, who already have to exercise hard in an effort to retain bone mineral density.

One of the most common methods of preserving food is canning. This is where the food has been heated for a couple of hours to a temperature that destroys microorganisms and then forms a vacuum seal as it cools. Some of the tastiest food that I ate in space was canned food, but the downside is that it is heavier and more bulky to dispose of. Only a small percentage of our food was canned.

Finally, there is 'bonus' food! This is an allocation of off-the-shelf items or specially prepared food that each astronaut can choose for their mission. Bonus food accounts for 10 per cent of our calorific intake, so astronauts must choose these meals wisely. I decided to donate a large portion of my bonus food towards running a competition for school children to design a healthy, nutritious and balanced meal for an astronaut, called 'The Great British Space Dinner'. The response was incredible, and the prize for the winning teams was to meet up with celebrity chef Heston Blumenthal to prepare seven dishes inspired by their creations. So I had a multi-Michelin-starred chef prepare my space food – not an entirely selfless decision, I admit!

There is a lot of tribal knowledge passed down from experienced astronauts as to what items work well, for bonus food. It is also a nice idea to have some food that you can share with your crewmates, to add a bit of cultural identity to social dining occasions. Since I have yet to see a can of fish and chips, my British contribution to space dining consisted of a bacon sarnie, chicken curry, sausage and mash, whisky-flavoured fudge, Yorkshire tea and Scottish shortbread.

Here's a small sample of the 'standard' food that we can choose from on the ISS:

Breakfast	Lunch	Dinner	Snacks/ Desserts	Beverages
Scrambled eggs	Split-pea soup	Barbecued beef brisket	Chocolate pudding cake	Coffee
Oatmeal	Chicken in salsa	Beef ravioli	Apricot cobbler	Tea
Granola	Pasta with prawns	Chicken with peanut sauce	Granola bar	Powdered milk
Sausage patties	Tuna-salad spread	Potato medley	Macadamia nuts	Cocoa
Dried fruit	Broccoli au gratin	Red beans and rice	Lemon-curd cake	Orange/ lemon/lime
Maple-top muffin	Tomatoes and aubergine	Creamed spinach	Butter biscuits	Strawberry breakfast drink

Q *Does food taste different in space?*

A This is a great question and one where the answer may vary, depending on who you talk to. I thought some of the food did taste different in space. I think the main reason is that we don't smell the food in space as much as we do on Earth – and so much of our eating experience relies on our sense of smell. This is partly due to the fact that there is no convection in weightlessness: hot air does not rise and cold air does not sink. Air inside the space station is moved around by ventilation fans and (much as in an aeroplane) this creates an artificial flow from the ceiling to the floor, taking smells away from our nose.

Of course, in space we have the perfect solution for this: eat upside down!

However, astronauts sometimes suffer from a reduced sense of smell, regardless of the lack of convection. In weightlessness, our body fluid shifts up towards our chest and head, causing astronauts to appear ' puffy-faced' and raising our intracranial pressure. Also the ISS is a dusty environment to work in, as particles float in the air and don't sink to the ground like they do on Earth. Both an increased intracranial pressure and a higher concentration of dust can cause inflammation to the nasal cavity lining, resulting in a stuffy nose and a subsequent reduced sense of smell.

And it's not just a lack of smell that affects our taste. Enjoying food is a multi-sensory experience and the ISS, with its clinical laboratory appearance, artificial white lighting, forced ventilation and remoteness from planet Earth is always going to struggle to provide an atmosphere for fine dining! I always enjoyed eating in the Russian segment. They had a couple of posters around the galley table, simple scenes of green fields, trees and spring flowers under a blue sky. It wasn't much, but it was home – and for me that made the food taste better.

Thankfully we were permitted a few condiments on board to liven up the food. We had salt and pepper in a liquid solution (it would be useless as grains, as they would simply float away) and items such as BBQ sauce, Tabasco and the all-important ketchup, for that bacon sarnie. In an effort to keep my salt intake low, I would often add some Tabasco to liven up an otherwise bland dish.

Q *What was your favourite space food?*

A There were several foods that I really enjoyed in space. Unsurprisingly, the food prepared by Heston Blumenthal and his team topped the list. In particular, canned Alaskan salmon with capers was an absolute favourite of mine. Ingredients such as capers, which explode in your mouth with an intense flavour, work really well in space. I would usually save those meals for the weekends, when I could enjoy eating them at a more leisurely pace than during our busy weekdays. However, when you are deprived of certain types of food, then even the most mundane-sounding snack can become a real treat.

I loved making myself a peanut-butter and jam sandwich as an afternoon treat. In space we don't have proper bread, so instead we used soft flour tortilla wraps, but it still tasted pretty good. And about once a week we could have a 'Maple-top muffin', which was a great way to start the day – especially with a bit of extra honey on top.

I think a huge part of the satisfaction that we get from eating has as much to do with the social setting as it does with the taste of the food itself. For that reason, some of my fondest space meals were those on Friday evenings, when the whole crew would come together to share a meal and relax at the end of a busy week. Usually we would all bring items from our bonus food and throw it together, in a truly international smorgasbord of goodies.

One of the things the crew always looks forward to is a visiting cargo spacecraft. Usually the ground crew pack a small supply of fresh fruit at the top, near the hatch. It never ceased to amaze me how powerful the smell of fresh oranges could be, after living in a stale atmosphere for so long. Fresh fruit was a huge treat for all the crew.

Okay, so here's an example of a pretty good day of eating in space (my nutritionist might disagree!):

Breakfast: Bacon sarnie* with ketchup, organic mashed-fruit pouch

Morning snack: Maple-top muffin . . . extra honey

Lunch: Sausage and mash*, broccoli au gratin, baked beans

Afternoon snack: Peanut-butter and jam tortilla-wrap 'sandwich'

Dinner: Alaskan salmon*, creamed spinach, potatoes au gratin, stewed apples for dessert*

* Bonus food provided by Heston.

Q *How did it feel when you first ate in space? Doesn't the food float back up?*

A Actually you can eat food just fine in space. Swallowing and digestion rely far more on muscles inside the body than on gravity. You can prove this to yourself by doing a handstand and eating a banana or something! Swallowing is a complex sequence that uses muscles in the tongue, pharynx and oesophagus to squeeze the food that you eat into your stomach. This process of muscle contractions is called 'peristalsis'. After swallowing, a ring of muscle fibres in the lower oesophagus closes, to stop the food and stomach acid coming back up again. If this process didn't work properly in space, then astronauts would suffer terrible heartburn caused by acid reflux.

Once safely inside your stomach, there are various valves and muscles that help digest your food and keep it moving in the right direction. However, after a meal you can definitely feel that your food sits more 'lightly' in your stomach, and I learnt the hard way that you shouldn't run on the treadmill for at least an hour or two after eating, if at all possible. Gravity is not required for the body to digest food, but it certainly helps.

Like every other system in our body, the digestive system will adapt to living in microgravity, and I think mine did a much better job of digesting food after a few days in space. Until then, the advice we had was to eat smaller amounts of food more frequently for those first few space dinners, until the body had a chance to figure out what was going on.

Q *Is it true that you lose your appetite in space?*

A This question is highly subjective. Some astronauts report having a much-reduced appetite in space, whilst others feel ravenous. Personally, my appetite reduced a little, but not by much. I still had a strong desire to eat food at mealtimes, but would often feel 'full up' after eating less food than I would normally consume on Earth. This was not helped by the fact that the portion sizes in our packets, cans and pouches of space food are quite small.

I lost about 5 kg in the first few weeks on board. This was a mixture of losing body fluid that you just don't need in space (this is the excess fluid that causes the 'puffy-face' look that astronauts often get) and not eating enough. Our diet is monitored regularly during the mission and, following my first nutritional assessment, I received the wonderful news that I needed to take on more calories. I interpreted that as a licence to eat chocolate-pudding cake or a similarly delicious dessert every evening! I regained most of my weight during the mission by a mixture of good diet and strength training. By the time I landed, I was a fraction under my pre-launch weight of 70 kg.

Q *What would happen if someone got sick or injured in space?*

A All astronauts are trained to a very high level in first aid. In addition, there are always at least two Crew Medical Officers (CMOs) on board, who can also deal with basic surgical and dental procedures, such as suturing or filling/removing teeth. Both Tim Kopra and I were trained CMOs. There is also a medicine cabinet on the space station containing everything from analgesic painkillers and antihistamines to sleep aids, antibiotics and local anaesthetics. Most cases of sickness and injury are not urgent or life-threatening. On the ISS this will usually allow time for the crew to have a detailed consultation with the flight surgeons on the ground, prior to deciding how best to treat the patient.

If we developed a serious illness, such as appendicitis, the situation would be assessed by the flight surgeons and a decision would have to be made as to whether it would be better to remain on board and treat it with antibiotics (most likely) or return to Earth. In this respect, the space station is less isolated than some places on Earth. For example, most of Antarctica's research stations are inaccessible during the winter months and a crew evacuation is not an option. On the ISS, our Soyuz spacecraft remains with us as a 'lifeboat' for the duration of our mission. Although it would have major consequences for the ISS programme, the option exists for astronauts to return to Earth in a matter of hours, in the event of a medical emergency. That said, the cramped Soyuz spacecraft is far from being an ideal ambulance, and re-entry is a punishing ride even for a healthy individual, let alone someone on the verge of peritonitis . . . but it is an option, nonetheless.

For more immediate circumstances, the ISS is also equipped with an Automated External Defibrillator (AED). During training each crew practises multiple emergency scenarios that require resuscitation techniques, including intraosseous infusion (injecting drugs directly into the bone marrow) during cardiopulmonary resuscitation (CPR). In weightlessness, it is not so easy to perform CPR! First, it's necessary to restrain the patient securely on a stretcher, and then the resuscitator needs to use straps around their own lower back or knees, to prevent themselves floating off. Some crew members straddle the patient instead, or turn themselves upside down as if doing a handstand on the patient's chest, in order to push against the 'ceiling' during chest compressions.

However, astronauts are probably most at risk from lacerations, fractures and eye injuries. Careless zipping around in weightlessness, through hatchways and around corners would be to invite a clash of head against metal. Also, we can move items in space (such as 145 kg EVA spacesuits) much more easily than we could on Earth. Although these items are weightless, they still have mass and could build up dangerous momentum if not handled carefully, resulting in crushed bones. Astronauts often have to tilt experimental racks weighing hundreds of kilograms on Earth, remaining constantly vigilant not to trap any body parts in the process. Eyes are also vulnerable to picking up small foreign bodies that are floating, at the mercy of the ventilation system. We go to great lengths to ensure that any potential small metal shavings are contained, when working with certain tools and equipment.

Thankfully, a major medical emergency has never occurred on the International Space Station, and during our mission there were no injuries of any relevance.

Did you know?

- It is *not* standard procedure for astronauts to have their appendix removed prior to flight.

Q *What would happen if there was a fire on the space station?*

A During my mission there were a couple of 'fire' emergency warnings, but thankfully they turned out to be false alarms. However, we treat every situation as if it were a real emergency, of course. If there's a fire on the space station, the first thing the crew does is to warn the others, gather together in a safe area and then start working on the problem. Sometimes the fire may be very obvious, such as large visible flames or billowing smoke. Such a fire occurred on the Mir space station in 1997, when an oxygen canister ignited and fuelled an intense fire that damaged the station and put the crew's lives at risk. NASA astronaut Dr Jerry Linenger was on board at the time and described the fire as a 'raging blowtorch'.

Astronauts would probably rate a fire in space as one of their greatest fears. What made the Mir fire particularly dangerous was that the flame grew rapidly in size and intensity, reaching the far wall of the module and threatening to burn its way through the hull. If that had occurred, then the crew would undoubtedly have perished in minutes as the station's vital atmosphere escaped rapidly into space. Furthermore, the flame was blocking the path to one of the Soyuz spacecraft, leaving no option to evacuate. The crew eventually extinguished the fire and, despite thick smoke permeating the station, were able to clean the atmosphere and remain on board.

Of equal concern are small fires that are hard to find – perhaps there's just a smell of burning or a smoke-detector alarm with no other indications. The space station is a large place, with hundreds of panels covering electrical equipment. It's imperative for the crew to rapidly locate the fire before it spreads, remove the electrical power (electricity is the most likely source of ignition) and extinguish the flames.

Astronauts have procedures to deal with every possible case of fire, depending on where it has occurred and the severity of the situation. The crew will often split into teams to deal with a fire. For example, one pair may remain in the safe haven, communicating with the ground and controlling the space station via computers. Another pair will be the fire-fighting team, moving through the space station with breathing apparatus to locate and extinguish the fire, using carbon dioxide, water mist or foam fire extinguishers. The third pair can support the fire-fighting team, retrieving equipment and closing hatches to unaffected modules, preventing the spread of smoke contamination. As you can imagine, all this activity requires careful coordination, and astronauts spend many hours training as a crew and with their ground teams until their response to a fire becomes second nature.

The smoke-detectors also trigger an automatic response from the ISS to shut down all ventilation systems, so as not to feed more oxygen to the fire, and to reduce the spread of smoke throughout the station. The crew have special hand-held detectors that tell us if the air is contaminated with carbon monoxide and other harmful gases – these are used to reveal when it is safe to remove our breathing apparatus. There are also special air filters and equipment on board to help clean and scrub the atmosphere, following a fire, and return the ISS to full health. Only as a last resort would the crew consider evacuating the ISS in their Soyuz spacecraft.

Interestingly, the Soyuz spacecraft does not have any fire extinguishers. The way to fight a fire in the Soyuz is to close your helmet and depressurise the whole spacecraft: no oxygen = no fire!

Q *How fast is the Internet in space?*

A Accessing the Internet on the ISS for personal use was painfully, painfully slow. For those readers who remember what dial-up was like . . . well, it was not that good! But hey, even the fact that I am answering a question about how quickly astronauts can browse the Internet in space is remarkable. Internet has been available on the ISS since January 2010, when NASA astronaut TJ Creamer sent the first tweet from space: 'Hello Twitterverse! We r now LIVE tweeting from the International Space Station – the 1st live tweet from Space! :) More soon, send your ?s.' Space-to-Earth communications have certainly come a long way from the early days of having only scratchy voice-calls when passing over ground-stations.

Just to be clear, the space station has very fast data connectivity to Earth. This is provided by a network of tracking and data-relay satellites high above the space station, in a geosynchronous orbit. Quite rightly, this communication link is primarily used for monitoring and commanding the ISS, uploading the data required to run scientific experiments and downloading the results. The high-speed data is not there for astronauts to catch up on Twitter during the evenings! Having said that, I would have loved a faster Internet connection occasionally, to be able to use applications such as Google Earth, as we were very limited in Earth observation tools on the ISS. In fact the crew often had to make do with handing around a paper *Rand McNally World Atlas* to try and identify ground features or towns that we had photographed earlier in the day.

The Internet speed varied, depending on how much bandwidth was allocated for crew use. Sometimes it would take more than a minute to open up a single web page, and at other times (at best) perhaps five to ten seconds, but it was never fast enough even to consider streaming video. The Internet usually improved slightly in the evenings, when the science payloads did not demand such high bandwidth and more could be allocated for personal use.

Q *Do you have Wi-Fi on the space station?*

A Yes! The ISS has Wi-Fi provided by a couple of BelAir Wireless Access Points in the US segment. This was only used to connect devices on the operations network, though, not for personal Internet use. However, it did mean that we could use iPads to access our procedures, daily schedule and other apps, which greatly improved crew efficiency. We would frequently have to visit several places around the space station prior to a task, just to gather tools and equipment, and then stow everything back in its correct place at the end. I really enjoyed the freedom of being able to nip around the space station with an iPad Velcroed to my knee, with all the information I needed close at hand.

Q *How did you use Twitter and Facebook from space?*

A Our crew quarters contained two laptops. One was connected to the 'operations network'. This computer gave us access to all the vital data required to conduct our daily work on the ISS, such as schedules, procedures, email and a myriad of other tools and applications. However, in order to provide a firewall to protect the operational data, we used a second laptop connected to a 'crew support network' to access the Internet and for 'live' posting on Twitter and Facebook.

In fact this second computer was not actually connected to the Internet, but instead was connected via remote desktop to another computer in Houston that accessed the Internet – a simple but smart way to avoid security problems on the ISS. Live posting and viewing comments on Twitter and Facebook were a great way to feel connected with what was going on back on Earth, but it was very time-consuming.

To post a photograph on Twitter, I would first have to upload it to my 'operations network' computer in my crew quarter. Then I would email the photo to my private Internet-based email account back on Earth. Next, using the 'crew support network' computer, I would have to access my private email account, save the picture to the remote desktop in Houston, then log into Twitter and finally I was able to upload and post the photo . . . phew! That whole process was never faster than five minutes.

A much more efficient method for using Twitter and Facebook was to email the pictures and posts directly to my support team at the European Space Agency. Communicating our experiences and sharing images of Earth are an important part of our job, and social media is an extremely useful tool for doing this. ESA had a small team who did a brilliant job of managing all my photographs, videos and social-media posts. Like most things we do in space, we're not doing it alone – it's all about teamwork. To that end, most astronauts spend very little time on social media whilst in space – the trick is to try and be as efficient as possible at all times, and that means using help from the ground when necessary.

Q *What exercises do you do to keep fit in space? – Aasiya, from Mellor Community Primary School in Leicester*

A Keeping fit in space is really important, not simply to be able to function effectively in weightlessness, but in order to cope with gravity again once the mission is over. The problem is that the human body is extremely good at adapting to new environments. Left to its own devices, our body would attempt to morph into the perfect being for living in space. This would make the return to Earth – or, in the future, a landing on the Moon or Mars – particularly punishing. To that end, the exercises that astronauts do in space focus on attempting to counter some of the negative effects caused by microgravity: namely, a reduction in muscle mass and strength, bone mineral density and cardiovascular fitness.

To stay fit and healthy on board the ISS, astronauts use a weight-training device called the Advanced Resistive Exercise Device (ARED), a treadmill (called T2) and a bicycle machine (called CEVIS). The ARED is a bit like a multi-gym that uses two piston-driven vacuum cylinders to provide up to 270 kg of resistive load (real weights would be useless in weightlessness, of course). This allows us to stimulate all the major muscle groups using a number of different exercises, such as squat thrusts, heel raises, shoulder press, bench press, sit-ups, upright row, bicep curls, and so on. An added bonus of ARED is that it is located just above the Cupola window, so when taking short breaks in between exercises the crew can float down and enjoy a rest with a perfect view.

The T2 treadmill uses a harness and bungee system to mimic body weight and keep astronauts attached to the treadmill. The tension in the bungee system can be modified by simply adding or removing metal hooks that adjust the length of the bungee. Most astronauts aim to target about 70 per cent of body weight when running, although this varies, depending on the type of training required each day. The treadmill can be used in a powered mode, like most running machines, or a passive (free-wheeling) mode, which is much harder as the astronaut has to self-power the treadmill against a fair amount of resistance. T2 is a great device to counter cardiovascular, muscular and bone deconditioning as it is both load-bearing and high-intensity.

The third exercise device is CEVIS. This is the bicycle machine and is primarily used for cardiovascular conditioning. The great thing about cycling in space is that you don't need a seat – it simply adds unnecessary weight and bulk. Instead, astronauts clip their bike shoes into the pedals, hold onto the handrails to stabilise themselves and pedal away.

When astronauts first arrive on board the ISS, there is a period of acclimatisation required for each of these exercise devices. Certainly, the T2 harness takes a while to get used to; and care has to be taken with ARED, due to the high loads combined with potentially unstable body movements in weightlessness, which can cause injury. Usually astronauts aim to continually increase the loading and intensity of their exercise regime throughout their mission, in order to get the best results and return to Earth in good shape.

I always enjoyed exercising in space, partly because I like to keep fit on Earth, too, but also because it was a chance to let the brain relax! Working on the ISS requires a high degree of attention to detail and a large amount of reading detailed procedures, both of which require intensive cognitive functioning. There was nothing better than putting on some good music or listening to an interesting podcast whilst having a tough workout.

Q *Was it hard running the London Marathon in space?*

A I think running a marathon anywhere is hard! Whilst I was in space, Eddie Izzard ran a staggering 27 marathons in 27 days across South Africa to raise money for Sport Relief. I was delighted to have the opportunity to call Eddie from the space station, the night before his final double marathon, to congratulate him on his incredible achievement. He was certainly a source of inspiration for my upcoming race, although my few hours on the T2 treadmill one Sunday morning paled in comparison to his endeavour.

I first ran the London Marathon in 1999, at the sprightly age of 27, completing it in 3 hours and 15 minutes. My lasting memory of that day was the overwhelming atmosphere of camaraderie, fun and support from the crowd throughout the entire course. To that end, I was delighted on the morning of my marathon challenge in space when Mission Control live-streamed the BBC's coverage of the race to the space station, so that I could watch it as I ran and feel part of the event on the planet below. I wasn't planning on beating any records, since my training regime had been limited by being in space, and running on the T2 treadmill involves some challenges. I thought about four hours would be a respectable time, and my training team back at the European Astronaut Centre had put a plan together to achieve that goal.

However, as I got to the halfway mark, my shoulders and waist began to hurt from the harness that was pulling me down on the treadmill. I was running with about 70 per cent body weight, so my legs were having an easier time than the runners back on Earth. Having said that, in order to run wearing the T2 harness system, you have to develop a slightly lolloping gait, which is unnatural and adds extra effort. I started to speed up slightly, to spare my shoulders from unnecessary pain. By the time I got to the 18-mile point the harness had become my nemesis. I knew that if I was going to finish the race, then it had to be over in the shortest possible time, so I sped up even more. My training team, monitoring the event back in Cologne and unaware of my pain, thought I was putting in a strong finishing performance. I was never more grateful to have the live TV coverage than in those final few miles, drawing encouragement from the crowds and the thousands of other runners (including some of my own support team, who were running the race wearing fabric astronaut spacesuits!). As it turned out, I completed the run in 3 hours

and 35 minutes. The relief at finally removing the harness and being able to float in weightlessness was monumental.

So yes, the marathon was quite hard, and I nursed my bruises for three days before I could put the harness back on and go for another run! However, it was also one of the highlights of the mission for me, and something I was proud to have taken part in.

Q *Mine is a bit of a daft question maybe, but . . . when I watched you run the London Marathon, I wondered what happened to the sweat you produced? I'm assuming you'd sweat normally, so did it float around in droplets or stay stuck on you and make you hotter, rather than helping you to cool down? – Caroline Mallender*

A This is not a daft question at all! When running, I thought that the sweat would form droplets on my skin and remain in place, without having the effect of gravity to pull the droplets downwards. This was indeed the case for my arms and legs. However, what was really interesting was observing the sweat on my face and head. The motion of running caused the droplets to coalesce into a much larger bubble of sweat that migrated to the top of my head. Every 20 minutes or so I would feel this bubble wobbling around in my hair and have to towel my head dry. I found the space station much warmer than I would have liked for exercising. I have always enjoyed running in a cool, damp environment (give me some good old British drizzle any day, as the perfect running weather!). As such, I probably sweated more than usual while running in the 21-degree warmth of the space station, so it was always important to drink plenty of water in order to rehydrate after exercise.

Q *What did you pack to go to space?*

A Astronauts start packing for space a long time before they launch, which requires a lot of forethought, planning and coordination. The majority of items that we will need in space (for example, bonus food, clothing, toiletries and a small selection of personal equipment, such as a multi-tool, torch and stationery) are chosen about 18 months prior to

launch. These items are packed and distributed on various cargo resupply spacecraft, to ensure there will be sufficient resources on board by the time the crew arrive. Other personal items that need to be flown ahead of time include exercise equipment, such as our T2 running harness, cycling shoes and running trainers.

After the essentials have been decided, astronauts are given a personal allowance – about the size of two shoe boxes – within which they can pack some personal effects. Usually astronauts choose a variety of items that they may wish to fly with for friends, family, charities or organisations with which they are affiliated. This is where forward planning really comes into the equation. I knew that I wanted to run the London Marathon and so I needed a running shirt for the 2016 race, well over a year in advance. I'm also a huge fan of rugby, so an England rugby shirt was packed, in preparation for the RBS Six Nations Championship. Other T-shirts included those of charities such as Help for Heroes, Raleigh International and the Prince's Trust. Then there were a number of items that were going to be used for the many educational outreach programmes that were planned for the mission, such as Mission X, Astro Pi, Rocket Science, Unlimited Space Agency, Astro Academy, and so on. I packed flags, too, to celebrate St George's Day, St Patrick's Day, St Andrew's Day and St David's Day, for England, Ireland, Scotland and Wales. I would hang them up in the European science laboratory and do a small video message to each country when it was their national day.

Finally, being British, I decided that I should be prepared for a more formal occasion and so I chose to fly a tuxedo T-shirt, just in case. As it turned out, this was perfect attire for when I was asked to present Adele with her Global Success BRIT Award from space. Having packed all that into my bag, there wasn't much room left, although I did squeeze in some photographs of friends and family to decorate my crew quarter with. In addition, my ever-thoughtful wife had cut a small corner off the blankets that my two boys, Thomas and Oliver (then six and four years old), used to sleep with, for me to take with me.

We were also given a small (1.5 kg) allocation by the Russians to fly with us in the Soyuz spacecraft. This was extremely useful, as the items could be decided upon much closer to launch. In this allotment I chose to take some personal effects that I wanted to fly to space, in addition to Helen Sharman's copy of Gagarin's *Road to the Stars*. We were also granted a small allowance in the Soyuz for any personal medication that we wished to have with us during our six-month mission, although there are already extensive medical supplies on board the ISS. And, most importantly, our EVA spacesuit gloves actually travel with us in the spacecraft, too. The remainder of the spacesuit resides on the ISS and can be modularly constructed to give a tailored fit, but the gloves are individually sized and usually travel with us in the Soyuz.

Q *What was the funniest moment in space?*

A Some of the strongest memories of space are not the incredible views of planet Earth or the liberating feelings of weightlessness, but the experiences that you share with your crewmates. In that respect, I consider myself extremely fortunate to have spent my time on the ISS with such an outstanding crew. The ISS commander often sets the tone for the working regime, and when I arrived that was Scott Kelly, who was already nine months into a year-long mission to the ISS. Scott is just the most brilliant guy to fly with: hard-working, sharp, ruthlessly efficient and generous to a fault. He's also got a great sense of humour.

Somehow Scott had managed to get a gorilla suit delivered to the ISS, and very few people knew about it. When he told me quietly about the gorilla suit I thought he was joking, or that perhaps it was just a face-mask, because you couldn't possibly get a gorilla suit into space, could you? Later that day I was proved wrong as I came face-to-face with a fully clad gorilla in Node 3. Scott's plan was to hide in Tim Kopra's crew quarter, then burst out and give him a scare. Scott asked me to go and tell Tim that he needed to make a call to Mission Control, knowing that the easiest place to do this was from our crew quarters.

So, with Scott hiding in position, I duly found Tim and told him that our Flight Director had asked him to call down. Of course he went straight to his crew quarter, only to be confronted with a hairy gorilla bursting out – completely unexpected and very amusing to watch. That was probably the funniest moment . . . at least the funniest one I can tell you about!

Quick-fire round:

Q *What kind of watch do astronauts wear?*

A Currently ESA astronauts are issued with the Omega Speedmaster X-33 Skywalker, which was developed in collaboration with ESA astronauts and features several functions that are ideally suited for spaceflight, such as multiple alarms (extra-loud, to overcome the background noise!). Astronauts can choose to wear pretty much any watch they like, provided it passes NASA's safety requirements. Certain watch batteries are not permitted, nor are watch faces that would shatter into pieces if struck, such as sapphire crystal. This would clearly present a serious risk of eye injury in a weightless environment and so durable, shatterproof materials such as hesalite crystal are used instead.

Q *What was your most essential item/tool on board?*

A I always carried a small torch and a Leatherman multi-tool on me. Both items I used several times a day. Often we would have to search for items in dark corners of the space station, so the torch always came in handy.

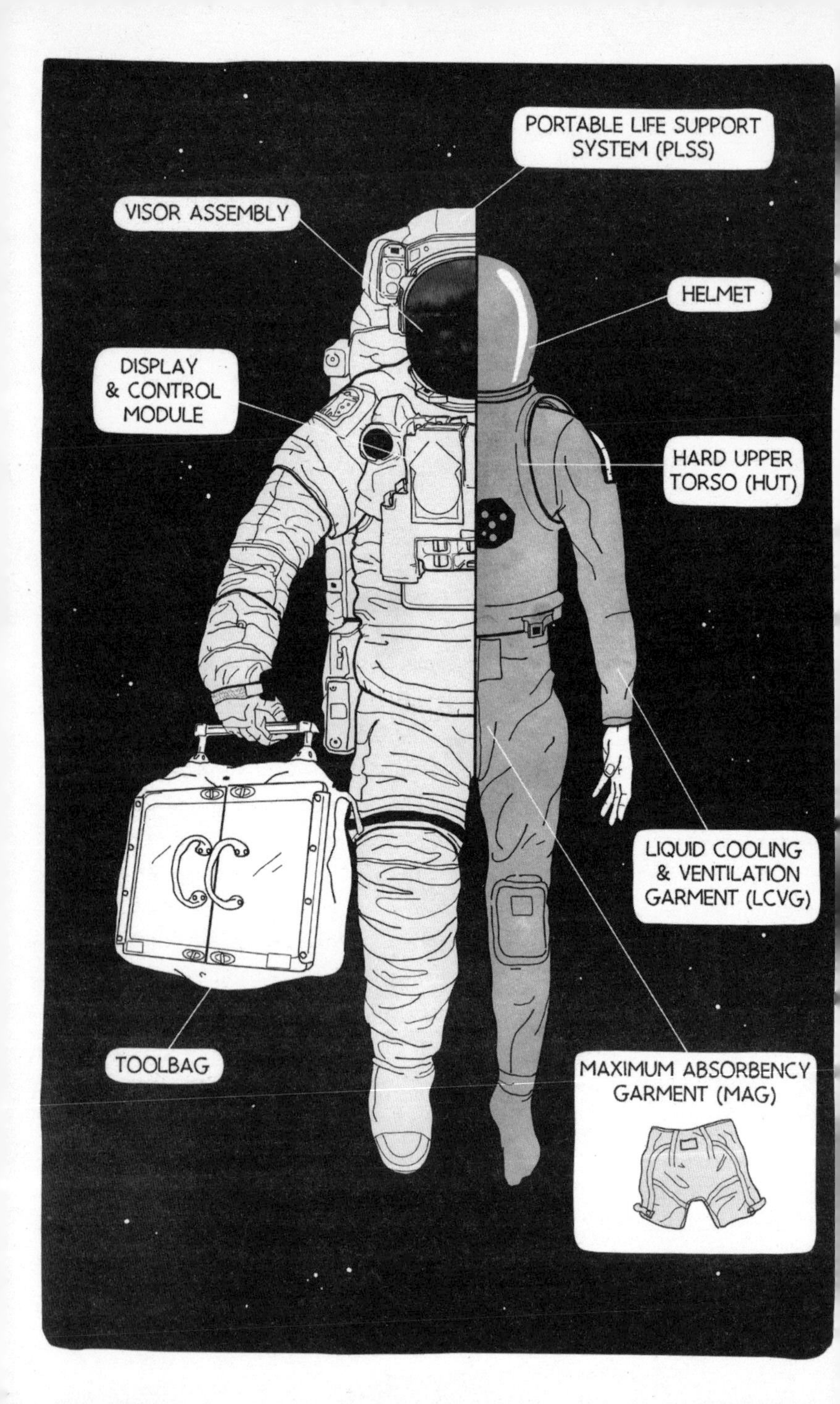
PORTABLE LIFE SUPPORT SYSTEM (PLSS)
VISOR ASSEMBLY
HELMET
DISPLAY & CONTROL MODULE
HARD UPPER TORSO (HUT)
LIQUID COOLING & VENTILATION GARMENT (LCVG)
TOOLBAG
MAXIMUM ABSORBENCY GARMENT (MAG)

SPACEWALKING

Q *What do you remember as your most amazing experience on the International Space Station? – Steph Webb*

A At 12.55 p.m. on Friday 15 January 2016, my fellow astronaut Tim Kopra and I got the message from Mission Control to exit the Quest Airlock. Carrying a bag of tools each and a replacement voltage regulator (a box roughly the size of a small fridge), we set off to repair a faulty solar panel – a would-be routine task back on Earth for any electrician. The only difference? We were setting foot into the vacuum of space. Leaving the safe haven of the International Space Station, we were about to enter an environment where temperatures are so extreme they can fluctuate from 200 to -200 degrees Celsius between sunlight and shade; where night turns to day in just 45 minutes; and where at any moment you could be hit by a hurtling micrometeorite. Worse still, you could lose your grip and find yourself floating in space.

My first ever extravehicular activity (EVA) – or spacewalk, as it's more commonly known – was the most vivid memory I have of life in space. It may only have lasted 4 hours and 43 minutes, but it was a day for which I had been preparing for years, and one I will never forget. Suspended high above the planet, it was surreal and thrilling to work in such a perilous place and to witness the widescreen beauty of Earth, as few in history have done. But before we delve into the detail of what it's really like, and what it takes to pull off something as audacious as walking in space, I want to take you back to a few days before I launched, to share a story of a remarkable pioneer in this field.

Q *When was the first-ever spacewalk?*

A On Monday 30 November 2015, Tim Kopra, Yuri Malenchenko and I gathered behind a long table displaying a splendid Russian breakfast buffet of various meats, cheeses, bread, fruit and pastries. It was just past 8 a.m. With a glass of vodka in hand, I was listening intently to a Russian gentleman give an extremely eloquent toast. The reason for this early-morning celebration was one of the many Russian traditions prior to launch – the breakfast ceremony in Star City. This is a last opportunity for friends and colleagues to bid farewell to the prime crew, before they head to Baikonur and into quarantine. The gentleman in question spoke with a strong voice, which defied his 81 years of age, and I was not alone in hanging on his every word. Silence had fallen over the crowded room as Alexey Leonov, twice Hero of the Soviet Union and the first person in history to conduct a spacewalk more than 50 years before, wished us good fortune on our upcoming mission.

On 18 March 1965, Leonov, a former fighter pilot in the Soviet Air Force, ventured out into the unknown from the relative safety of his Voskhod 2 spacecraft – an achievement that ranks in my mind with Yuri Gagarin's first orbit of Earth nearly four years previously, and another major coup for the Soviets in beating the Americans to the post. Despite Leonov's EVA lasting only 12 minutes and 9 seconds, it was both a pioneering and a perilous accomplishment. In the vacuum of space, his Berkut ('Golden Eagle') spacesuit had ballooned and stiffened to the point at which he could not pull himself back in, using the umbilical cord connected to his spacecraft. Inside his expanding suit, his fingers had slipped out of his gloves and his feet no longer reached into his boots. Not wishing to alert Mission Control, Leonov took action and did the only thing he could: reaching for his spacesuit regulator, he started to slowly depressurise his spacesuit. As he released the pressure in his suit, he risked starving his body of oxygen and suffering potentially fatal decompression sickness, but he reasoned that if he couldn't get back into the airlock, then he would be dead anyway. When he finally reached the small, inflatable fabric tube attached to the side of his spacecraft that served as an airlock, he already had pins and needles – the first symptoms of decompression sickness – and his core body temperature was soaring, due to the extreme physical effort. Leonov entered the airlock in the only way possible: head first. He was supposed to enter feet first. The airlock's narrow diameter had barely enough room for a spacesuit, let alone sufficient volume for someone to turn round in order to close the outer hatch. Soaked with sweat, unable to see through the perspiration and dangerously close to suffering heatstroke, Leonov somehow managed to manoeuvre himself so that he could close the hatch and return to his capsule.

As I shook the hand of this remarkable man I wondered if I would ever have the opportunity to follow in his footsteps and venture outside the International Space Station. As it turned out, I didn't have long to wait for that answer. Shortly after our mission commenced, Tim Kopra and I were assigned to the 192nd spacewalk on the International Space Station. I had been given the opportunity to fulfil a lifelong dream.

Did you know?

Here is a list of ten of the most experienced spacewalkers in history, who have spent the most cumulative time working in the vacuum of space.*

* The list is current as of July 2017.

Ranking	Astronaut	Space agency	Total EVAs (Extravehicular activities/ spacewalks)	Total time Hours: minutes
1	Anatoly Solovyev	RSA*	16	82:22
2	Michael Lopez-Alegria	NASA	10	67:40
3	Peggy Whitson	NASA	10	60:21
4	Jerry L. Ross	NASA	9	58:32
5	John M. Grunsfeld	NASA	8	58:30
6	Richard Mastracchio	NASA	9	53:04
7	Fyodor Yurchikhin	RSA	8	51:53
8	Sunita Williams	NASA	7	50:40
9	Steven L. Smith	NASA	7	49:48
10	Michael Fincke	NASA	9	48:37
* RSA (Roscosmos State Corporation for Space Activities) is the Russian space programme.				

Q *What was the best part of your spacewalk?*

A When Tim Kopra and I egressed the airlock on 15 January 2016, our main objective was to replace a failed Sequential Shunt Unit, or SSU, located at the base of one of the solar panels on the farthest starboard edge of the space station. The SSU receives coarse voltage from the solar panel and regulates it, so that the solar array operates at a constant voltage and load. With the SSU failed, the space station was down one-eighth of its electrical power, so it was an important job to restore the space station to full capability.

Tim and I had to get to the worksite in a safe, yet expeditious, manner. Timing was critical to this EVA. Since the SSU receives coarse voltage from the solar panel, there is no way of switching it off. The only way to change it out safely is to wait until the Sun goes down . . . no Sun = no solar-power generation. Our timings were based on being in position at the worksite and ready to change out the unit just before sunset. As it happened, Tim and I made good progress and we were in position ten minutes ahead of schedule. Rather than putting us to work on another task and risk something detracting from our main objective, Mission Control elected to have us wait in position until darkness. Being told to 'hang out' for ten minutes on a spacewalk and watch a sunset, whilst floating at the very edge of the space station, is unheard of. Having grabbed the opportunity to take some photos (including the obligatory spacewalk 'selfie'), I had about five minutes to soak up the view and reflect on my situation.

During my spacewalk by far the best part was a feeling of awe and reverence during those precious few minutes of 'hanging out' in space. As we crossed the terminator from day into night, it was like having a front-row seat in nature's own IMAX theatre. I had a similar feeling when I first looked out of the Cupola window, although spacewalking took this to another order of magnitude. I had the ability and freedom to turn in any direction, one minute marvelling at how fragile and beautiful Earth looked as it slipped gracefully into shadow, the next minute being intimidated by the vast blackness of space stretching into infinity. It certainly gave new meaning to the word 'perspective'. Unencumbered by the effects of gravity, not feeling the weight of my spacesuit or noticing the thin visor in front of my eyes, I felt completely detached – utterly removed from Earth, civilisation, the space station. I had the sensation of being a microscopic spectator in an immeasurably vast universe. It was, at the same time, the most astonishing and humbling experience of my life.

Q *Did you feel scared at any point?*

A Any activity where you're pushing technology and human performance to the limit of what is possible is incredibly exciting – and, yes, it comes with its fair share of apprehension. Any nerves that I had before my spacewalk were dealt with by ensuring that I was completely prepared for what I was about to embark upon. But there's still nothing worse than waiting, whether it's for an exam, a job interview or a spacewalk. I've always found that action brings a sense of calmness, and the moment I felt any apprehension disappear was when Tim opened the hatch to space. Night was approaching and the Sun was low on the horizon. I remember sunlight suddenly flooding into the depressurised airlock and thinking, 'Finally, time to get to work!'

No matter how relaxed you may feel about stepping into the vacuum of space, you still need to remain extremely vigilant against something going wrong. The one thing I've noticed, when sharing experiences of spacewalking with other astronauts, is a common understanding of just how extreme that environment is. One astronaut described the feeling of danger as 'palpable'. This is not to say that astronauts are thrill-seekers or adrenaline junkies, but rarely in my life have I ever felt in such a precarious situation as during those few hours spent outside the space station. There's no denying the exhilaration of a spacewalk!

Q *What was it like to wear the first Union Flag on a spacewalk?*

A Astronauts are embedded in a large but extremely close-knit team that supports human spaceflight operations around the world. The space station is a truly international affair, and as a crew member on board the ISS, you feel like you are representing a global effort in science and exploration. However, there are also times when you are reminded of the significance, honour and impact of representing your home nation in this worldwide endeavour. This was never more apparent to me than when I had the privilege to wear the Union Flag on my spacesuit as I stepped out into space. I think Scott Kelly summed up the importance of that moment perfectly when he said, 'Hey, Tim, it's really cool seeing that Union Jack go outside. It's explored all over the world. Now it's explored space.' It was one of the proudest moments of my life.

When the spacewalk was over, and Tim and I were safely back inside the space station, we got straight to work tidying tools and equipment and debriefing on the day's events. It wasn't until much later that evening that I realised the magnitude of the support I had received back home. My crew support team had sent up a few of the messages of encouragement that had been flooding in during the day. There was even a tweet from Sir Paul McCartney: 'Good luck – we're all watching, no pressure! Wishing you a happy stroll outdoors in the universe.'

It was both overwhelming and extremely humbling to realise that what had been a personal dream, and an achievement that I had worked long and hard for, had also meant so much to so many people back home. I went to sleep that night a very proud Briton – with the Union Flag from my spacesuit Velcroed to the wall in my crew quarters.

Q *I've heard that astronauts can get 'the bends' in space. How is that possible, and what would you do to treat it?*

A To answer this question we need to talk a bit about the pressure inside our spacesuit. When venturing into the vacuum of space, if we didn't have a pressurised suit we would lose consciousness in about 15 seconds and die shortly thereafter. This is because pressure keeps dissolved gases in solution. There is a lot of dissolved gas in our bodies, mainly nitrogen and oxygen in our blood and skin tissue. Living under the weight (pressure) of Earth's atmosphere keeps that gas in solution. Remove that pressure, though, and dangerous bubbles will begin to form. At best, this can cause itching of the skin and pain in the joints; at worst, bubbles are transported to the brain, causing paralysis and death. This is known as decompression sickness or 'the bends'.

So we need some pressure inside the suit to keep us alive. However, if we inflate the suit to sea-level pressure on the inside, versus a vacuum on the outside, we would look like the Michelin man from those tyre-company adverts. It would be extremely hard to move physically in the suit against such a large pressure differential, not to mention the increased structural strength required from the suit to withstand repeated pressurisation cycles. Instead, the pressure inside the suit is maintained at about 4.3 psi (one-third of an atmosphere), which is a compromise between safety and flexibility. That's enough pressure to keep the body's dissolved gases in solution, but it's pretty low nonetheless. This low pressure allows you to bend your arms and fingers more easily against the suit's rigidity, but you're getting to the stage where dissolved nitrogen in the body can start to form minute bubbles and cause the bends.

In order to mitigate this risk, astronauts prepare for a spacewalk by flushing as much nitrogen out of the body as possible beforehand. We 'pre-breathe' 100 per cent oxygen soon after waking on the morning of a spacewalk, wearing breathing masks connected via long hoses to the space station's oxygen supply. Then we depressurise the airlock to 10.2 psi whilst donning our spacesuits, and shortly thereafter we conduct a strict protocol of light exercises (kicking our legs) for 50 minutes whilst suited. All of these measures help to reduce the risk of an astronaut getting 'the bends' during a spacewalk. However, astronauts are also trained about the risks of decompression illness and remain constantly vigilant, monitoring themselves for any signs or symptoms throughout a spacewalk.

If an astronaut did suffer from the bends, then – depending on the severity of the situation – we would first recover them back to the space station and, whilst keeping them inside their spacesuit, would increase the pressure in the suit to above the space station's normal atmosphere. This would enable the bubbles of gas to dissolve into the body once again. Under the careful guidance of the flight surgeon, we would then slowly reduce the pressure to normal. In effect, we would use the spacesuit as a personal decompression chamber to treat the bends, in a similar manner to the way a patient with decompression illness would be dealt with on Earth.

Did you know?

- The pressure inside our spacesuit (4.3 psi) is equivalent to an altitude on Earth of more than 9,000 metres – that's lower than the pressure at the summit of Mount Everest (4.89 psi). However, astronauts breathe 100 per cent oxygen in the suit, which allows us to go to such low pressure and still remain mentally sharp.

Fact or fiction?

Would your blood boil in space, if you didn't wear a spacesuit?

Strictly speaking, no – the exposure of your body to the hard vacuum of space would certainly cause dissolved gas in your bloodstream to form bubbles, so if you could take a picture, it would look a little like a boiling liquid, but the plasma and cells wouldn't be 'boiling', in the strict sense of the term. Bear in mind that you wouldn't be around all that long to appreciate the difference, as lots of other bad (and terminal) things would be going on at the same time.

Q *Do you have your own spacesuit or do you share it with other astronauts?*

A There are two types of spacesuit. Astronauts wear a smaller, lighter and 'softer' spacesuit inside the spacecraft that takes them to and from the ISS. This spacesuit is designed to be comfortable when strapped into a seat, and usually relies on the spacecraft to provide oxygen or air to pressurise and ventilate the suit. Normally the suit will only be pressurised in the event of an emergency, when the spacecraft itself can no longer maintain pressure to protect the crew. In the case of the Soyuz spacecraft, this suit is called 'Sokol' (meaning Falcon).

Each Soyuz crew member has an individually 'made-to-measure' Sokol suit, as it is vital that the suit fits correctly in the small confines of the descent module. There are also adjustment straps on the arms, legs, chest and abdomen, to account for minor changes and personal comfort.

Prior to flying in space we have to endure a two-hour 'fit-check' in a fully pressurised Sokol suit, strapped into a Soyuz seat. Once pressurised, the suit expands slightly and becomes more uncomfortable. If the leg length is not perfect, then excess material can dig in behind your knees or on the tops of your ankles. What can seem like a minor annoyance after five minutes can grow to become a major distraction after an hour or more, causing extreme discomfort. That's the purpose of the fit-check – to rectify those problems before strapping yourself into a rocket. I was very fortunate (or perhaps it's a result of being only 5 feet 8 inches tall) and found the suit so comfy that I slept through most of the fit-check.

In contrast, the spacesuit used for conducting spacewalks is akin to a mini-space station. The US suit is called an Extravehicular Mobility Unit (EMU) and weighs in at a whopping 145 kg. The EMU has to keep an astronaut alive for up to eight hours (sometimes more) in the harsh environment of space. We don't have our own EMU suit (that would be way too expensive and inefficient). Instead we share the majority of the suits with other astronauts, but have the capability to tailor it to achieve a near-perfect fit. The 'Hard Upper Torso' comes in four sizes: small, medium, large and extra-large, although the small size has never been flown in space. Boot sizes are just medium or large, with the option of boot inserts to fill any gaps. However, there are about 50 or 60 sizes of glove to choose from. This is because it is so important to try and optimise fidelity and dexterity in your hand movements during a spacewalk. The pressurised gloves feel clumsy enough as it is, but our suit engineers do an excellent job of trying to perfect the fit. If you want to know what it's like working in EVA gloves, then try tying your shoelaces whilst wearing a pair of oven mitts! Our arm and leg lengths can be adjusted with metal spacing rings and fabric fasteners, and the helmet is one-size-fits-all.

Astronauts spend many hours wearing the EMU in the pool during training, and on each occasion the suit engineers offer advice as to how to optimise the overall fit. Finally, prior to our mission we wear our 'Class I' EMU suit (this is the real thing – spaceflight-certified hardware, as opposed to the training suits we use in the pool). In order to ensure a good suit fit, we need to replicate the conditions of space as best we can, and for that reason we use a vacuum chamber at the Johnson Space Center in Houston to test everything, prior to a spacewalk.

During that vacuum-chamber run I remember three very cool things happening. The first was that my spacesuit felt fantastic – it was a perfect fit, which was a huge relief. Second, a small bowl of water had been placed on the floor of the chamber as a demonstration of what happens to water in a vacuum. As the pressure inside the chamber reduced, the water suddenly began to boil quite violently, then froze and sublimated (a process whereby a solid turns directly into a gas without first becoming a liquid) – something I had never observed before. Third, I had taken a feather and a coin into the chamber with me. When it was at vacuum, I placed them both on a piece of card and released them simultaneously. Watching the coin drop to the floor was unremarkable – it appeared no different from usual. However, watching the feather plummet to the floor at the same rate as the coin was truly incredible. I knew it was going to happen, since there was no air to slow down the feather's fall to the ground – but all the same, it was really weird to see a feather drop to the ground like a brick!

Did you know?

- The Russians have a separate spacesuit for spacewalking called the 'Orlan' (Sea Eagle). Similar to the EMU, it provides about seven hours of life support and weighs 120 kg. The suit operates to a higher pressure of 5.8 psi, making it even harder to work in, but with less risk of the bends.

- As a general rule, Russian cosmonauts use the Orlan spacesuit to conduct spacewalks on the Russian segment of the ISS and US, Canadian, Japanese and European astronauts use the EMU to conduct spacewalks on the US segment of the ISS. However, there have been several exceptions to this rule. I was fortunate to be trained in the use of both suits, but wore the EMU for my spacewalk.

Q *How are the routes planned for spacewalks on the ISS?*

A We call the routes on a spacewalk 'translation paths'. As you can imagine, a lot of thought goes into choosing the best routes. Usually each spacewalk will have been performed many times in the pool, prior to astronauts executing it for real. During these pool 'development runs', the job of the EVA team on the ground (which includes experienced astronauts) is to make the spacewalk as safe and efficient as possible. The planned translation paths will be practised, modified and studied for hours by the development team, before being sent up to the crew on board the ISS.

Many factors are taken into account in the route-planning, such as 'no-touch zones', hazardous areas, emergency crew-recovery options, optimal efficiency and the level of difficulty. Some parts of the space station are relatively easy to navigate, with plenty of hand-holds, and are free from obstructions. Other areas are not so accommodating, and traversing them is like trying to plan a severe rock-climbing pitch with an overhang! Needless to say, we try and stay away from such areas, unless absolutely necessary. We're also very cognisant of our safety tethers – on a spacewalk we are like spiders, weaving a trail of thin steel wire wherever we go. It's important to plan our translation paths so that this wire doesn't become tangled or create an obstacle for a crewmate.

Ultimately, the crew conducting the spacewalk have the option to

change the route, in agreement with the EVA team. I spent several hours planning my translation paths, using a combination of laptop-based virtual reality (VR) software and simply looking out of the windows. The VR software was a brilliant tool for planning, but nothing quite beats studying the real thing, and I found that looking out of the space-station windows and observing (as well as I could) the actual route I would be taking paid huge dividends during the spacewalk. My approach to preparing for a spacewalk was similar to planning a flying sortie during the early days of my aviation training. The night before an important 'check-ride' I would sit in my room and visualise the entire sortie, from start to finish, in excruciating detail: where I was going; what controls I would be manipulating; what radio calls I'd need to make; which actions to take in the event of an emergency; and so on. By adopting a similar approach with a spacewalk, I felt as if I had already completed it, even before stepping outside the space station. As Canadian astronaut Chris Hadfield says, 'An astronaut who doesn't sweat the small stuff is a dead astronaut.'

Q *When you're out on a spacewalk, how do you go to the toilet?*

A On the day of a spacewalk astronauts can spend 12 hours or more in the spacesuit, with no access to the loo. For this reason, astronauts wear adult nappies under their cooling suits, in case the need should arise. These are the same 'Maximum Absorbency Garments' or MAGs that we wear during a Soyuz launch. Usually the EVA crew will get up at about 6.30 a.m. on the morning of the spacewalk, spend a few minutes on personal hygiene and then carefully place electrodes on our chests so that the flight surgeons will be able to monitor our heart rates during the spacewalk. Soon after this we don our MAGs, long underwear and Liquid Cooling and Ventilation Garments (LCVGs). Then we start breathing 100 per cent oxygen via a face mask, as a precaution against getting 'the bends'. About an hour after this we have a final trip to the loo (whilst still breathing pure oxygen), prior to donning our spacesuits. The process of flushing nitrogen from our bodies takes a while, so by the time we finally step out into space, we've already been wearing our

22: After more than ten hours inside the cramped Soyuz spacecraft, and following an eventful docking, we were glad to be safely on board the ISS and greeted by the welcoming faces of Mikhail Kornienko, Sergey Volkov and Scott Kelly.

23–25: Collecting data for science is a huge part of life on the International Space Station. Some of the studies that I found most interesting were 'life-science' experiments – whether taking blood samples, testing for airway inflammation or investigating muscle atrophy – the research was always fascinating.

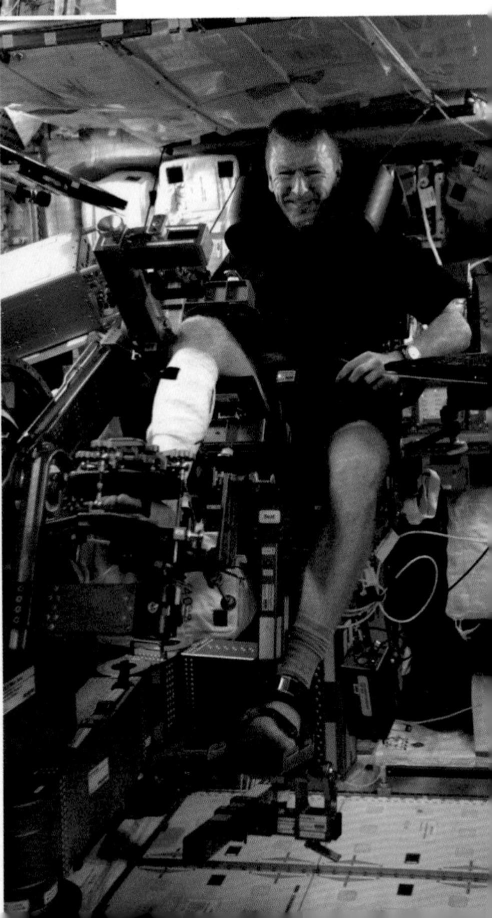

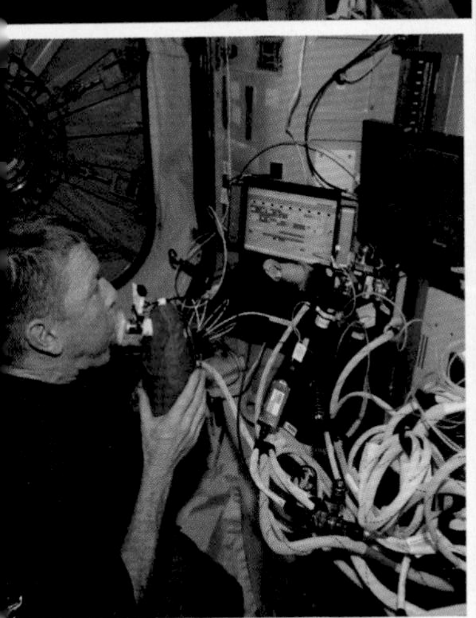

26: We would always look forward to a resupply spacecraft arriving... with the welcome addition of fresh fruit to an otherwise monotonous diet.

27: The International Space Station barber shop consisted of a self-inflicted dodgy haircut once every two weeks. The tube connected to the clippers is attached to a vacuum cleaner to prevent the loose hair from floating around the space station.

28: Charged particles from the Sun penetrate the Earth's magnetic field, colliding with atoms and molecules in our atmosphere. The result is a spectacular aurora, dancing eerily above the Earth's poles. The camera really doesn't do it justice...

:9: One of my favourite photographs has to be this rare shot of Antarctica, which is so far outh of the ISS orbit that it's extremely unusual to get a clear shot.

:0: Planets and stars can be seen clearly from the ISS. Here, Venus is rising a few moments efore the Sun. From space, above the turbulence of Earth's atmosphere, the light from stars nd planets doesn't 'twinkle' like it often does on Earth.

Orbiting the planet 16 times a day, it is not long before you become very familiar with even the most remote and rugged places on Earth. There are so many places that I would now love to visit...

31: The Andes, South America.

32: The volcanoes of the Kamchatka Peninsula, in the Russian Far East.

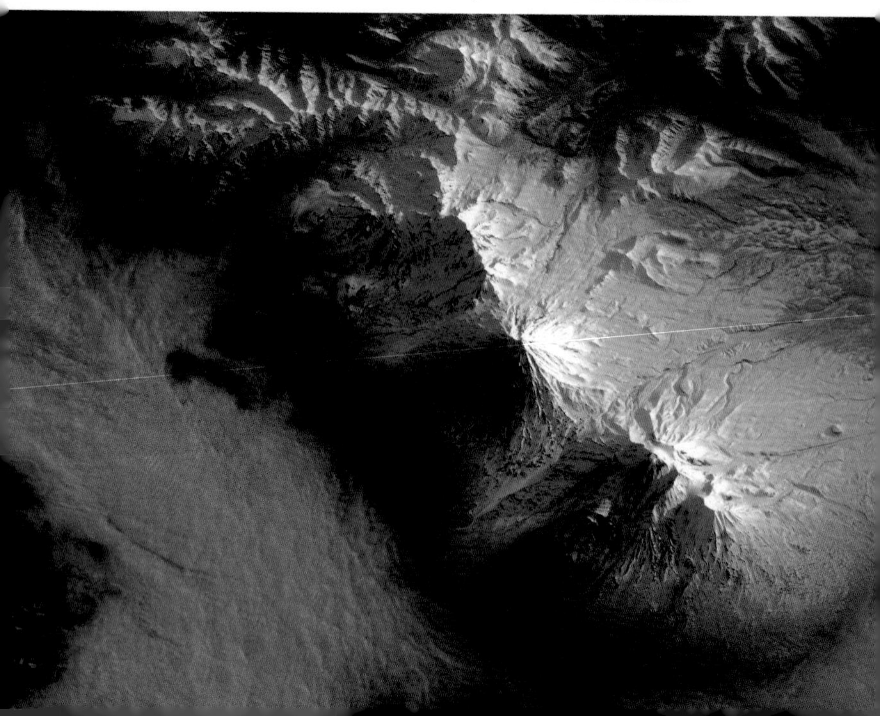

33: Nam Co Lake, China. In Mongolian it is known as *Tengur nuur*, meaning ‘Heavenly Lake’.

34: Coast Mountains, British Columbia.

35: Lake Alakol and the Almaty province, Kazakhstan.

36: It takes a couple of weeks to get used to sleeping in space. It's not as simple as you would think and never quite as satisfying as it is on Earth – collapsing into bed and resting your head on a pillow.

37: Running any marathon is difficult. In space, the harness that keeps you attached to the treadmill creates an additional challenge. The relief I felt at finally removing it and floating again in weightlessness was monumental.

38–39: My first ever spacewalk was one of the most exhilarating experiences during my time on the ISS. It may only have lasted 4 hours and 43 minutes, but it was a day for which I had been preparing for years, and one I will never forget.

40: Life on the ISS is extremely busy, but on Sundays we usually had some free time to call family and friends back home, whilst watching the world go by!

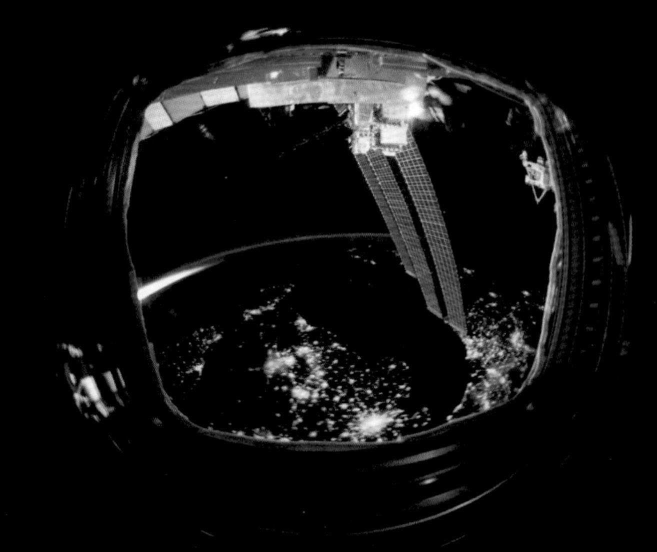

41: After almost six months in space my spine had elongated and I had grown nearly 2 inches. With only three weeks until undocking, it was time for a seat fit-check in our Soyuz spacecraft.

42: It's a pretty wild ride returning to Earth. From entering the atmosphere at 99.8 km to the parachute opening at 10.8 km took an exhilarating 8 minutes and 17 seconds. Here, the soft-landing thrusters are firing just prior to touchdown.

43: After living in space for so long, Earth's gravity can feel punishing. In particular, astronauts are often plagued with dizziness, nausea and vertigo in the first 48 hours after landing. But nothing beats the fresh smell of Earth after six months on the ISS.

spacesuits for about five hours. We then have around six hours of spacewalking ahead of us – and although I never needed to use my MAG, I was certainly glad to be wearing one, just in case!

Did you know?

- One of the worst things you can do, if you need to pee – whether on a spacewalk or not – is to hold it in for a long time. It can not only be very painful, but can also cause bladder complications. Unfortunately this has happened during previous missions in space, and on one occasion an astronaut required medical assistance for seven days, in order to urinate. Not only can this severely impact on your ability to perform certain tasks, but it can also introduce a risk of infection and even jeopardise the mission. So, if you need to go for a pee during a spacewalk, then it's far better to let it go than hold it in!

Q *In scuba diving there is a syndrome known as 'fear of surfacing', where divers don't want to come up. Did you ever feel like this on your spacewalk?*

A I've read about this phenomenon and have also enjoyed some spectacular dives myself, where the temptation to go deeper and stay longer is very strong. These are the circumstances when discipline and training come to the fore. In aviation, and particularly in test flying, we would agree on critical decision points and 'knock-it-off' criteria before the sortie. That way, there was no discussion if the temptation arose to push the limits further than had been briefed – everyone knew when it was time to come home.

With regard to spacewalking, it is certainly one of the most extreme, exhilarating and incredible situations someone can find themselves in, and I would gladly have remained outside the space station for a few hours more. Sure, it's extremely tempting to stay out for longer – after all, those precious moments will undoubtedly remain the most memorable in my life. And I certainly wouldn't be the first astronaut longing for a little more outdoor time. During the very first US spacewalk, NASA astronaut Ed White had to be ordered back in, after 23 minutes of spacewalking. He joked, 'This is fun. But I'm coming.' After a few more minutes of stalling, he finally squeezed back into the Gemini capsule and said to his commander, Jim McDivitt, 'It's the saddest moment of my life.'

However, the circumstances of our spacewalk made it very clear to everyone that it was time to return to the airlock. Tim Kopra's spacesuit had developed a fault, and water was beginning to enter his helmet via the ventilation duct at the back of his head. At that point in the spacewalk Tim and I had been working on separate tasks, but we were not too far from each other. By the time I was able to see inside Tim's helmet, there was already a golfball-sized globule of water lurking ominously on the front of his visor. Tim and I had been training together for well over two years. As a flown Space Shuttle crewmember and experienced spacewalker, I had learnt an enormous amount from this calm, modest and exceptionally talented astronaut and I couldn't think of anyone I would rather entrust my life to on a spacewalk. As I reported the quantity of water in Tim's helmet to Mission Control, we both understood the seriousness of the situation. No one needed reminding of the dangers of water in a spacesuit helmet.

A similar situation had occurred in 2013, during my ESA classmate Luca Parmitano's second spacewalk. Luca's situation deteriorated rapidly, with an even larger quantity of water sticking to his face and covering his eyes and nose. Shortly after this, the water in his headset caused his communications to fail and he could no longer hear instructions, talk to Mission Control or his spacewalking crewmate, Chris Cassidy. If that wasn't bad enough, the space station was rapidly approaching darkness.

In space, twilight does not linger as it does on Earth, waiting for the Sun to dip gracefully below the horizon. Instead, travelling at Mach 25, one minute it is bright daylight and the next minute it is pitch-black.

Now blind, deaf and dumb, and wondering if his next breath would result in a lungful of water, Luca was relying on the small pull from his retractable safety tether to guide him back to the sanctuary of the airlock. With help from Chris, Luca eventually made it safely back inside and, when they finally pulled off his helmet, about 1.5 litres of water were inside – that's a lot of water to have floating in the small confines of a helmet. That incident was a close call, and it was one of the most serious emergencies to have occurred on the ISS to date. So when Tim Kopra's helmet started filling with water during our spacewalk, it wasn't long before Mission Control told us, 'Guys, you can start opening your cuff checklist to page seven; we are in a Terminate case.' It was time to 'come up for air'. Fortunately, despite the premature end to our mission, we had already completed the main objective of restoring the ISS back up to full power, and the spacewalk had been declared a success.

Did you know?

- Following ESA astronaut Luca Parmitano's 'water in the helmet' incident, a NASA report identified the primary culprit as a blocked water separator, which resulted in water spilling into the ventilation loop. Since then, several changes have been made to procedures, equipment and training to mitigate the risk of a subsequent occurrence. As part of these changes, two modifications were incorporated into the spacesuit. First, astronauts now affix a snorkel that leads from their helmet down to the waist area of the suit. This allows them to breathe from an uncontaminated part of the suit, in the event that water enters the helmet. The second change is to place a

Helmet Absorption Pad, or HAP (a bit like a modified nappy), at the back of the helmet. This will catch any water coming into the helmet from the ventilation loop. Astronauts can check the 'feel' of this HAP against the back of their head during a spacewalk, in order to try and detect water ingress. So a snorkel and a nappy – I love how sometimes the simplest solutions can be applied to the most complex of problems.

Q *Why do astronauts train underwater for spacewalks?*

A A lot of time and effort is spent training for spacewalks in an underwater environment because water gives us 'neutral buoyancy', which is a little bit like weightlessness. Our spacesuits are usually filled with oxygen, or just with air when we're training in the pool. We need this air to breathe and to pressurise the suit against the vacuum of space. Left to its own devices, a spacesuit filled with air will float on the water like a big balloon. However, when training for a spacewalk we need to go underwater, so weights are placed around the spacesuit to make it neutrally buoyant and evenly balanced. This means that it will neither float nor sink, but will hopefully remain perfectly in one spot, when released at a specific depth. Achieving this delicate balance is called a 'weigh-out' and is an acquired skill that depends on the expertise of our support divers. A good weigh-out makes all the difference, when training for six hours underwater – if it's not accurate, you'll be continually fighting the suit's tendency to want to float or sink, expending vast amounts of energy and rapidly becoming fatigued.

So neutral buoyancy enables astronauts to practise moving and working outside the space station for hours at a time on a simulated spacewalk, suspended underwater, as a close simulation of floating in space. Of course, neutral buoyancy is not the same as weightlessness and there are some important differences. In space, *everything* moves, if it is not strapped down. The tiniest of forces will send objects spinning off, tumbling into space, never to be seen again. It's very easy to start things moving in space, whereas in water it's much harder. Next time you're underwater try pushing something away from you (preferably an object, not another person!) – it's quite hard to do, because water is viscous and causes drag. When we're training underwater we have to remember that in space things will start to move much more easily than they do in the pool.

You also have to be extremely careful with 'heavy' objects in space. Something that is heavy on Earth will of course appear weightless in space, but it still has mass. A 100-kg object on Earth still has a mass of 100 kg in space – its *weight* has changed, but not its mass. This means that objects in space can easily build up momentum (defined as an object's mass multiplied by velocity). When training in the pool, if you move a heavy object too quickly and it starts getting out of control, the viscosity and drag of the water will help to slow it down. On a spacewalk, there is nothing to stop this momentum except your own strength or an impact with another part of the space station, which is never popular with Mission Control.

Furthermore, even though we're neutrally buoyant underwater, we still feel the effects of gravity, so although we can rotate ourselves upside down with relative ease, the blood still rushes to our head and all of our body weight rests on our shoulders inside the spacesuit. This is really uncomfortable – it can become painful after a few minutes and make it hard to equalise the pressure in our ears. In space, of course, there is no up or down and astronauts can turn themselves in any direction without feeling the effect of gravity.

Q *What's the most physically challenging thing you've done as an astronaut?*

A That's an interesting question. On the one hand, training events such as ESA's seven-day caving expedition, NASA's 12-day underwater NEEMO mission or enduring freezing winter temperatures during survival training in Russia have all presented physical challenges. After all, the point of this training is to put you under pressure and give you the confidence and skills to manage difficult situations. On the other hand, I sometimes think that the punishing two-and-a-half-year training schedule alone, with multiple international flights to Russia, Germany, Canada, Japan and the US, was a physical endurance test in its own right!

However, if I had to choose one thing that has been the most physically challenging, then it would be EVA training. Spacewalking is hard work – seriously hard work – and so training for a spacewalk needs to be, too. This is partly due to the physical effort involved. Inside the suit, astronauts are fighting the suit's pressure, with every small motion of our arms, shoulders and fingers draining valuable energy and raising our heart rates. Spacewalking is also extremely demanding mentally, and maintaining such a high level of concentration for hours, where the slightest mistake can have the direst consequences, is exhausting.

I relished the challenge of EVA training and, for me, it was the most enjoyable part of my training as an astronaut. Perhaps this was due to the fact that I was able to approach spacewalk training in the same way

I would approach a test-flying sortie, in terms of planning, preparation and execution. Our training runs in the pool would usually last for six hours, and for that entire period we would probably maintain the same heart rate as if we were running at a slow jog. It's strange to think that something that can look so easy when watching a spacewalk, such as an astronaut plugging in an electrical connector or simply moving around the space station, may be taking an enormous amount of effort inside the suit. We're like ducks in fast water – serene on the surface, but paddling like heck underneath!

The important thing when wearing the spacesuit is to try and maintain a steady workload and try not to sweat. This is sometimes difficult, as there will undoubtedly be periods of greater and lesser workload over six hours. There were a couple of times during my training when sweat got into my eyes – not something you want to happen in space, as the salt makes your eyes start to stream, but the liquid has nowhere to go. Something similar happened to Canadian astronaut Chris Hadfield during his first spacewalk, when one eye became contaminated. All the fluid that built up in response filled his eye socket, then spilled over into his other eye. Now blinded, it took Chris about 30 precious minutes of his spacewalking time to resolve the issue until he could see properly again.

Did you know?

- Despite being called a 'spacewalk', astronauts rarely use their legs during an EVA. The majority of the work is done by the upper body, in particular the shoulders, forearms, wrists and fingers. Occasionally astronauts may use a foot restraint to give them added stability at a worksite. This involves securing your feet onto a metal foot-plate, but otherwise the legs are pretty redundant during a spacewalk.

Q *Is it really true that Velcro was invented for astronauts to scratch their nose with, whilst in their spacesuit? This is what my grandpa told me, and I don't know whether to believe him or not . . . If it is true, did you have some Velcro inside your helmet? – Solomon, aged six*

A Well, Solomon, that sounds like a good use for Velcro, and I'm sure I would have believed your grandpa, if he'd told me that story. But I did some digging, and it seems there's a myth out there that NASA invented Velcro. In fact a Swiss electrical engineer named George de Mestral came up with the idea for Velcro in the 1940s. He was inspired after walking his dog and wondering if the burrs that clung to his clothing (and dog hair) could be turned into something useful. De Mestral finally patented his idea in 1955, a few years before humans had even ventured into space.

However, Velcro is used extensively in spaceflight due to its lightweight, strong adhesive qualities, and it can be manufactured out of materials that are flame-retardant and can tolerate thermal extremes. We do use Velcro inside our helmet and it has something to do with scratching our nose, so your grandpa is not too far off the mark there. It's a foam 'valsalva' device that we stick, using Velcro, low down on the inside of our visor and it helps us to clear our ears as the pressure changes inside the spacesuit. It also doubles as a great nose-scratcher!

Q *Was there anything that surprised you on your spacewalk, which really caught your eye? – Anonymous*

A There was one point during my spacewalk when I was returning to the airlock with the failed SSU that Tim Kopra and I had replaced. I had to descend along a thin metal spur that connects the main truss section to the airlock. A lot of the time during a spacewalk is spent working close to large pieces of structure, and this gives you a sense of security. However, halfway along this exposed spur I looked down and noticed Australia passing beneath me and I felt a sudden tinge of vertigo. Instinctively I tightened my grip on the handrail. It made me smile, because I had been 'out the door' for well over an hour by that stage, but seeing an entire continent 400 km beneath my feet caught me by surprise. NASA astronaut Chris Cassidy had advised me, if that ever happened, to wiggle my toes and it would make me relax my grip . . . and it worked!

Q *What would happen if you fell off the space station?*

A Falling off the space station is most astronauts' worst nightmare. In the opening scenes of the 2013 movie *Gravity*, Sandra Bullock's character finds herself detached from the Space Shuttle, tumbling uncontrollably through space, left to the mercy of the laws of physics. Under these circumstances, an astronaut would eventually die, most likely asphyxiated after several hours, as the spacesuit's ability to scrub carbon dioxide from the atmosphere slowly failed or the battery power ran out. It is no surprise, then, that we go to extraordinary lengths to ensure that astronauts do not go tumbling off into space to suffer a lingering death.

It's actually surprisingly easy to fall off the space station. The spacesuit gloves that we wear are bulky and cumbersome. The palms are coated with a special rubber material that gives adequate grip, but the gloves' thickness does not allow much fidelity or 'feel'. It's difficult to know how hard you need to grip something. To begin with, most astronauts end up holding on too tight, but with practice you learn to relax your grip and think like a rock-climber. The outside of the space station is covered with handrails and other structures that we can hold on to. However, there are also plenty of places that are dangerous to touch, either because they are sharp and could cut open a glove or because we might cause damage to the space station. So the first line of defence against falling off boils down to good planning, preparation and training: know where you are going.

I studied for hours the routes that I would follow during my spacewalk, analysing the reach between each hand-hold, the best body positions needed to span difficult gaps and planning alternative routes, in case things didn't work out. In addition to memorising planned routes and worksites, you also need to be able to think on your feet, in case there's a requirement to work in an area you've not prepared for. Hours of underwater training for spacewalks give you the skills and confidence needed for this.

The second line of defence is a mantra drilled into all rookie astronauts: 'You stop, you drop' – meaning that whenever you stop moving, the first thing you do is attach yourself to the space station using a short 'local' tether (about 1 metre long). Astronauts often have to let go of the space station with both hands, in order to use tools and equipment, work on a task, and so on. If you allowed yourself to become distracted, it would be all too easy to forget to first attach your local tether, and then let go and drift off into space.

Our third line of defence is a safety tether. This is like an oversized fishing reel, with one end anchored to the space station and the other end attached to our spacesuit, dispensing a thin steel wire on a spring-loaded retractable reel wherever we go (it was the small tug from this reel that allowed Luca to find his way back to the airlock, when his helmet filled with water during a spacewalk). However, this thin steel wire is a double-edged sword, and astronauts must remain constantly vigilant not to get tangled up in their own or their crewmate's safety tether. When planning a spacewalk, astronauts take special care to try and use separate routes or to think of a strategy that avoids potential entanglement.

Finally, if all else fails, our spacesuits have a built-in jet-pack, called SAFER. But as much fun as it sounds to go scooting around space with a jet-pack, I don't know of any astronaut who would relish the thought of having to use this last line of defence.

Q *What happens if you drop something on a spacewalk?*

A Unfortunately items have been dropped in the past during a spacewalk, and sometimes hardware failures have caused items to become detached from the space station. The utterly demoralising fact about weightlessness is that even if something is slowly drifting just out of fingertip reach, there is no getting it back. Crews have had to watch valuable tools and panels drift inexorably into the black void, and there is precious little they can do about it.

In March 2017, I was watching two very experienced astronauts and spacewalkers placing four large protective panels over a docking port. With three panels in place, they turned to locate the final panel, only to see it drifting unsecured beneath the ISS on its leisurely way back to Earth. In theory, this should never happen. As is the case in many mishaps, it is seldom one person's action or a single equipment failure that results in the error, but a series of unfortunate events that enable a mistake to be made. In aviation we call this the 'Swiss cheese' model: when all the holes line up, allowing an error to propagate and an accident to occur.

To try and protect against losing items in space, astronauts uphold a strict tether protocol during a spacewalk. Absolutely everything is attached to something else, in a chain that eventually leads back to the space station. Take a socket, for example. It could be attached to a driver, which is tethered to a tool caddy that is tethered to the inside of a toolbox. The toolbox is tethered to the astronaut, who is tethered to the space station. Whenever astronauts hand over tools to one another or attach things to the space station, we *always* 'make before break' – meaning that first we place a new tether onto the equipment and give it a 'pull-test' before releasing the old tether. This is much like the via-ferrata style of protection used in alpine climbing. Furthermore, each tether hook requires two separate actions to open it, in an attempt to prevent inadvertent release. The downside to using all these tethers is that life can quickly become a tangled mess, unless you are extremely organised and disciplined.

Part of the preparation for a spacewalk is to consider in excruciating detail what tools are required for what task, and in which order. Astronauts will organise their toolboxes and equipment so as to optimise efficiency, minimising the number of tethers required and then placing tools in exactly the right sequence, in an attempt to reduce the risk of entanglement. Of course in microgravity everything floats around, and occasionally, despite the best efforts of the crew, life on a spacewalk gets dogged by a tether snag or tangle.

On rare occasions the tether protocol breaks down. Perhaps an item was flown into space without equipment being secured correctly, or perhaps the crew made a tether connection but the gate on the hook didn't close fully, resulting in the tether inadvertently coming loose again. Whatever the reason, if something is 'dropped' in space and not tethered, then it is usually lost for ever. All the crew can do is report to the ground as accurately as possible the velocity and direction of the missing equipment. Getting a good video image is also a priority. By doing this, the experts in Mission Control can start identifying and tracking the lost item immediately, to determine whether it is going to present a danger to the space station.

No sooner does an item get lost in space than it becomes yet another piece of space debris – one that is now in an almost identical orbit to the space station. It is highly likely that the object will continue to move away from the space station and will eventually burn up in Earth's atmosphere. However, just like the space station, that lost item is travelling at Mach 25 and is orbiting the planet every hour and a half, meaning that Mission Control now have 90 minutes to be certain that it will not present a threat and come back to collide with the station on the next orbit.

Q *Can you eat anything during a spacewalk?*

A Unfortunately, during a spacewalk we only have the option of remaining well hydrated – there's nothing to eat. Before getting into their spacesuit, every astronaut fills a drinking bag that can hold about a litre of water. This is ordinary space-station water (that is, room-temperature recycled urine), with no added salts, caffeine or energy products. The drink bag is then Velcroed inside the front of our spacesuit, so that we feel it against our chests once we have squeezed our way into the suit's 'Hard Upper Torso' or HUT.

The drink bag has a small straw that extends towards the astronaut's chin, poking into the top of the helmet, with a rubber mouthpiece that allows you to bite it open and suck up the water. During training, each astronaut tries out various positions for the straw and mouthpiece until they find their perfect solution. It's important to get it right – too high and it will get caught in your chinstrap or annoy you every time you move your head. Too low and you risk not being able to reach it at all. During training I made a rookie mistake one time by taking a big gulp of water and then releasing the straw from my mouth without due care and attention. Water flicked up all over the inside of my visor, and for the remaining few hours of my simulated spacewalk I had to try and peer around these annoying droplets that wouldn't move – much to the amusement of the support divers around me.

NASA are currently looking into the possibility of adding protein or carbohydrate supplements to our drink bags, but as yet astronauts simply drink water during a spacewalk. For this reason it's important to have a good meal the night before a spacewalk and a decent breakfast the following morning, before putting the spacesuit on. Just like a runner preparing for a marathon, loading the body with carbohydrates will help to provide a reservoir of energy that the body can tap into during the spacewalk.

Q *How do you stay warm when it is so cold in space?*

A Objects in space can fluctuate between being extremely hot and extremely cold – and that's a serious challenge, not only for our spacesuits, but also for anything else that has to endure such harsh thermal extremes. When talking about temperature in space, it's important to realise that we are not talking about air temperature, as we do on Earth. That's because there is no air in the vacuum of space. Without air there is no convection, and so our spacesuits have to deal with heat transferred by conduction (for example, when astronauts touch part of the space station) or by radiation.

The hot plasma of the Sun emits photons of energy, some of which are absorbed by objects in space, warming them up. At the same time, photons are constantly radiating away from any object that has a temperature above absolute zero. This perpetual balance of receiving and emitting photons will determine an object's temperature. For instance, a piece of bare metal exposed to direct sunlight outside the space station can get as hot as 260 degrees Celsius. However, an object in the shade can be colder than minus 100 degrees Celsius. The space station comprises many elements, each with different thermal properties and experiencing different levels of exposure to the Sun. During a spacewalk, astronauts will inevitably touch objects at varying temperatures and their gloves will have to deal with these thermal extremes. Another challenge is that astronauts can easily float between working in the shadow to working in the light from the Sun, causing the suit to experience rapid changes in thermal exposure.

To compensate for these temperature extremes, the spacesuit is designed with multiple layers of material that provide insulation from heat loss (from your body) or heat gain (from the Sun). Called

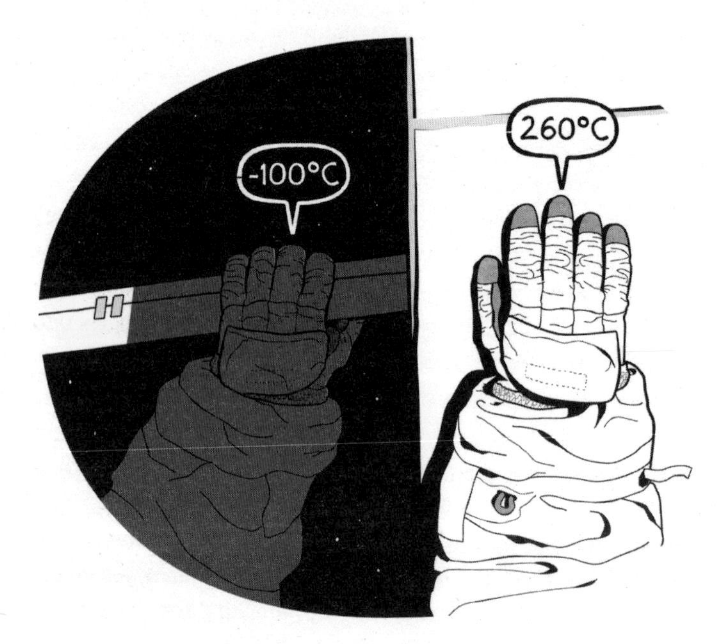

'multi-layer insulation' or MLI, this material is in fact used extensively outside the space station in an attempt to suppress temperature variations and protect sensitive pieces of equipment. The spacesuit is so good at insulating against both the heat and the cold that during my spacewalk I only needed to adjust my temperature twice in nearly five hours.

So, how do you stay warm out in space? Well, we rely on the body's own heat. During a spacewalk you're working pretty hard, generating plenty of body heat and trying not to sweat. This body heat is sufficient to keep you warm, with the exception of your fingers, which can sometimes get a bit cold. To protect against this, our gloves have electrical heaters that warm the fingertips when activated. Each time we approach sunset, Mission Control warn us, so that we can switch on our glove heaters.

With the body generating all this heat and the spacesuit trapping it so effectively, an equally tough challenge is . . .

Q *How do you keep cool in space?*

A Well, to stay cool, astronauts wear a special suit called a Liquid Cooling and Ventilation Garment (LCVG) under their spacesuits. The LCVG is worn close to the body and incorporates coils of very small-diameter clear-plastic tubing. Cool water is pumped through this tubing to chill the body, when needed. So where do we get the cool water from? That's the really clever part. Some of the spacesuit's water supply is passed through a 'sublimator' – a porous plate that exposes the water to the vacuum of space. This water freezes and then slowly sublimates into vapour, with the vapour vented off into space. This forms a kind of ice-pack that water can be passed through, to chill it and remove the heat generated by our bodies.

There is a metal dial on the front of our spacesuits called a Thermal Control Valve (TCV). The TCV mixes water that has been chilled with water that has bypassed the sublimator – a little like a shower unit mixing hot and cold water. This enables the astronaut to select the desired temperature of the water that is being pumped through the tubes close to their body. This system is so effective that it's actually very easy to cool off when working hard, but sometimes harder to warm back up again. For this reason, astronauts concentrate on maintaining a steady work rate during a spacewalk, trying to avoid temperature extremes or the need to adjust the cooling valve.

Fact or fiction?

In the movie Gravity, *Sandra Bullock is wearing just hot pants and a sleeveless top under her spacesuit – is this correct?*

No! Under our spacesuits we wear a nappy, long johns, a long-sleeved top and then the LCVG – not quite as sexy as hot pants and a sleeveless shirt, but way more practical.

Q *Is it hard to work in the dark, out in space?*

A During daylight the Sun is so bright that most of the time astronauts elect to have their gold-coated visors down during a spacewalk. Yes, we know this looks really cool, but actually the main reason for keeping our visors down is so that we are not blinded inadvertently by looking at that giant nuclear furnace called the Sun. As we approach sunset, Mission Control warn us to raise our visors, in preparation for a rapid transition into the shadow of Earth.

Spacewalking during the 'night' phase of our orbit can be very challenging. Although we have helmet lights (these are switched on continuously during a spacewalk), they only illuminate a small pool of light directly in front of us. When we are stationary, this radiance is sufficient for completing any tasks that we may be working on, and in fact at one point I was concentrating so intently on what I was doing that an entire night-pass went by – without me really noticing the dark at all. However, it becomes more challenging when you have to move to other parts of the space station in the dark. The ISS is a big place – during daylight it is easy to see where you are going and to orientate yourself. During night-time this becomes much more difficult, and the helmet lights only solve part of the puzzle as to where you are.

Mission Control will often switch on external lighting around the space station to assist astronauts on a spacewalk. This can be useful, but sometimes lights just seem to 'float' in the blackness of space and, if you change orientation, then trying to maintain spatial awareness of where you are and which direction you're supposed to be going is not as easy as it might sound. To help with this, especially in the case of an emergency, there are a few arrows painted on modules outside the space station showing the direction back to the airlock. Simple, yet effective.

Did you know?

- The reason our visors are coated in a thin layer of gold (they are also made of a polycarbonate plastic) is threefold. Firstly gold is a pliable material to work with. This property is important because it means it can be manufactured into a very thin transparent layer, which astronauts can still see through. Secondly, gold doesn't tarnish, by which I mean it doesn't corrode or rust. This is another vital property for the helmet's visor to possess, so that it does not lose its reflectance. And the final reason we use gold for our visors is that gold is very good at reflecting harmful infrared radiation, as well as other radiation from the Sun, which could otherwise permanently damage our eyes.

Q *What would happen if you were hit by space debris during a spacewalk?*

A Our spacesuit has 14 layers of material that protect the astronaut. Not all of this material is designed for thermal protection. Some of these layers are dedicated to pressure integrity, fire protection and defence against micrometeorites. In particular the outer layer of the suit consists of a puncture-resistant 'Thermal Micrometeoroid Garment', made partly from bulletproof material and designed to withstand impacts from small pieces of debris.

In addition, the chest and back are protected by the 'Hard Upper Torso' (HUT) and by the 'Portable Life Support System' (PLSS), both of which are built from hard materials and include a number of metal components. It's hard to be specific about whether a piece of space debris would puncture the suit or not, because it depends on the type of debris, the closure speed and the point of impact. Furthermore, although there is a lot of debris in low Earth orbit (both natural and man-made), space is a big place and astronauts are pretty tiny in comparison. When we assess the risk of an astronaut being hit by a piece of debris during a spacewalk, it falls into the category of potentially catastrophic, but *extremely* unlikely.

However, let's assume that our unlucky astronaut does get hit by a micrometeoroid. First, this is akin to being shot by an *exceptionally* high-velocity projectile. Since we are already travelling at Mach 25, it is highly likely that the combined velocity of a collision between astronaut and space debris will be many times the speed of sound. On impact, the suit will try to do its job, which is to dissipate the energy of the collision through multiple layers of material and prevent the debris from rupturing the pressurised bladder. If successful, it is possible that a minor strike will not even be noticed during a spacewalk and will only be realised later on, during an inspection of the suit.

If the debris does make it through the pressurised bladder, then the suit will begin to leak oxygen through the hole into space. Clearly, this is not a good situation, but it may not be catastrophic. If the hole is 6 mm or less in diameter, then oxygen from the two primary tanks will continue to maintain pressure inside the suit. The astronaut will receive an 'O2 USE HIGH' message, which will alert everyone to the fact that 'Houston, we have a problem'. As the primary tanks run low, two secondary oxygen tanks will take over. There will be more warning messages, and at this point the astronaut has around 30 minutes of oxygen remaining. That's not much time, but if there are no further complications, it should be sufficient to make it safely back to the airlock.

If the hole is greater than 6 mm, then the situation is a little bleaker. The maximum flow rate that the suit can provide is about 3.2 kg/h – not sufficient to maintain pressure in a suit with a big hole, regardless of how much oxygen remains in the tanks. In the event of a large leak, the pressure inside the suit is going to drop rapidly. At 3 psi (that's equivalent to nearly 12,200 metres altitude), the astronaut will receive a 'SUIT P EMERGENCY' message (the 'P' is for pressure) and soon after they

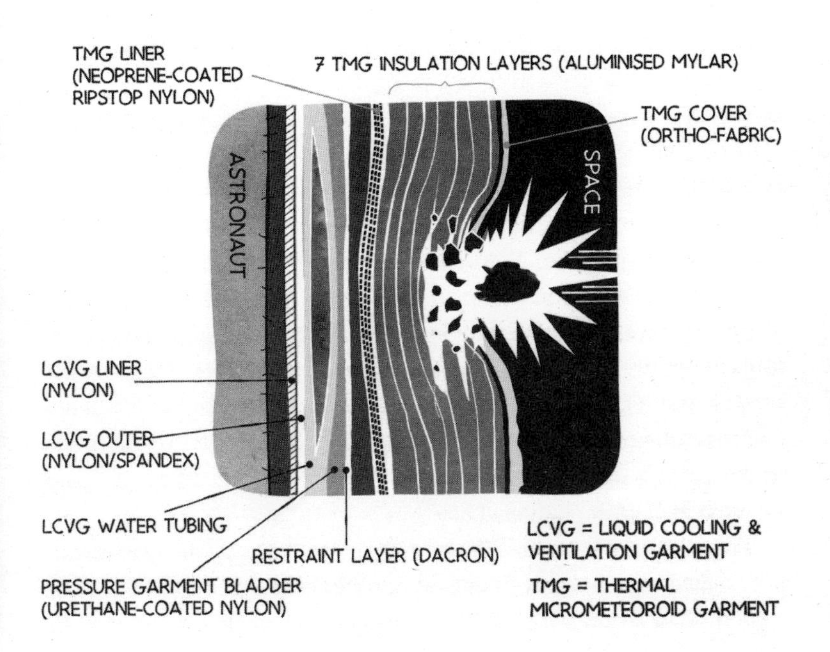

will begin to lose mental function. All of this assumes, of course, that the astronaut wasn't just killed outright by this high-velocity projectile penetrating the suit. Sorry – not the most cheerful answer!

Fact or fiction?

In the movie The Martian, *Matt Damon uses duct tape to cover a hole in his leaking helmet. Would this work?*

Quite possibly. The size of the hole in a leaking spacesuit can make the difference between life or death. Any attempt to cover the hole, reduce its size or patch it up is going to be effort well spent. Of course the tape would have to withstand the pressure inside the suit – but reducing the flow rate might buy you enough time to get to safety.

Q *Which astronaut has been your hero or has inspired you in your career?*

A There are many answers to this question that would seem an obvious choice. Who could possibly not be inspired by Yuri Gagarin, John Glenn, Alexey Leonov, Neil Armstrong and Valentina Tereshkova, to name but a few on the long list of revered men and women in the history of human spaceflight. However, to my mind one of the most inspiring figures, and an unsung hero of mine, is NASA astronaut Bruce McCandless. On 12 February 1984, McCandless became the first astronaut to conduct an untethered spacewalk, 'free-flying' with a jet-propelled backpack to a distance of 100 metres away from the safety of Space Shuttle *Challenger*'s payload bay.

To put that into perspective, everything we do today on a spacewalk is designed to prevent us from ever becoming detached from the space station. That's because to do so is dangerous – really dangerous. An accomplished US naval aviator with more than 5,000 hours' flight time, McCandless was selected for astronaut training during the Apollo era in 1966 and was CAPCOM, or Capsule Communicator, during the first lunar EVA of Apollo 11. McCandless had to wait 18 years to make his first spaceflight and during this time he worked, amongst other things, on the development of the Manned Manoeuvring Unit (MMU), the nitrogen-propelled backpack that he first flew – a truly groundbreaking event in the history of spacewalking.

I often wonder at the level of isolation McCandless must have felt, floating alone above the planet. During my spacewalk, whilst at the very farthest edge of the space station, it was exhilarating enough just to look over my right shoulder and see nothing but the black void of space. To have actually ventured out into that darkness with only a jet-pack and a leap of faith that nothing would go wrong with his equipment must have taken enormous courage.

Eventually McCandless logged more than 312 hours in space, including four hours of EVA MMU flight time. It goes to show that if you stay focused on your dreams, you can achieve anything.

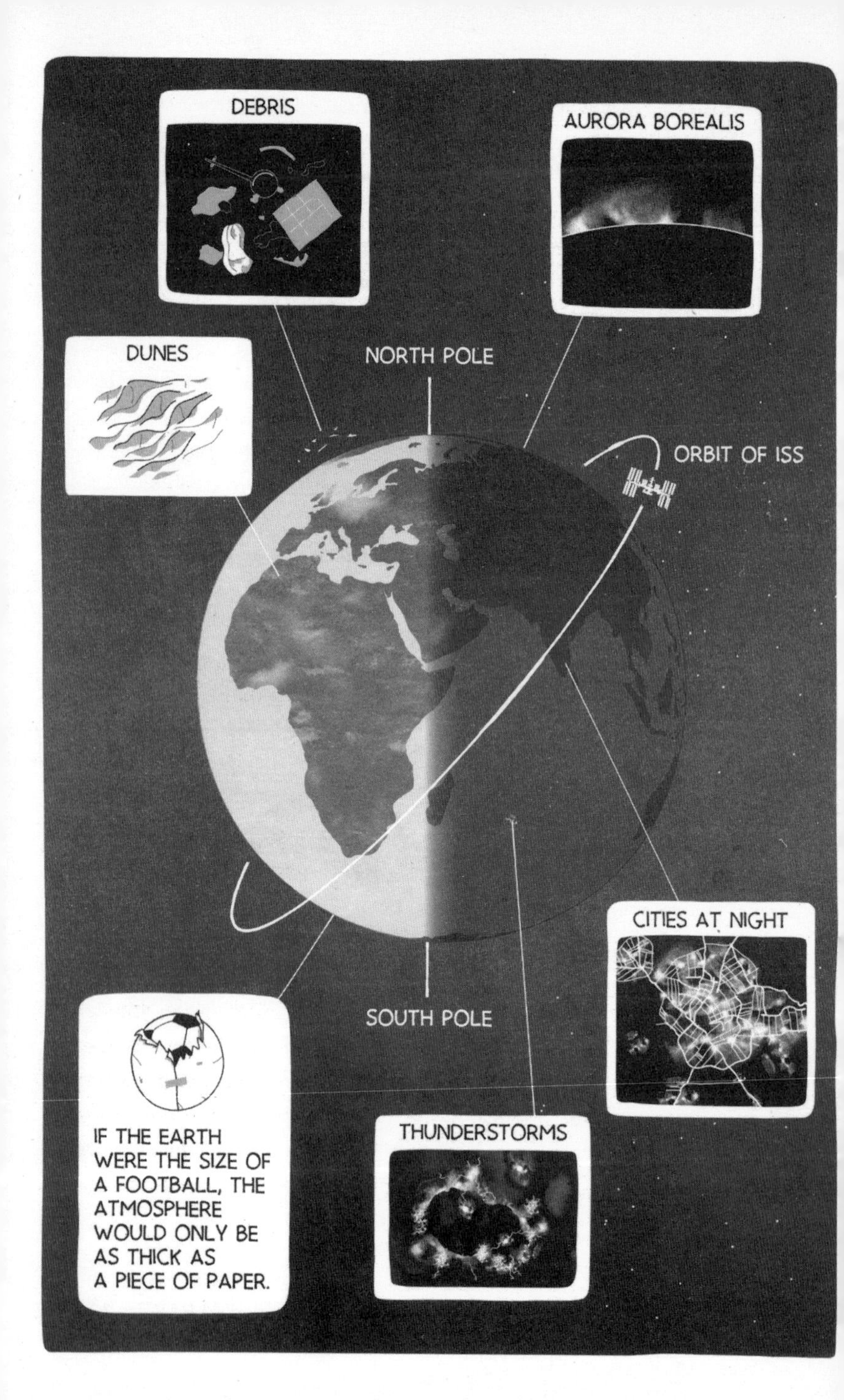
DEBRIS
AURORA BOREALIS
DUNES
NORTH POLE
ORBIT OF ISS
CITIES AT NIGHT
SOUTH POLE
IF THE EARTH WERE THE SIZE OF A FOOTBALL, THE ATMOSPHERE WOULD ONLY BE AS THICK AS A PIECE OF PAPER.
THUNDERSTORMS

EARTH AND SPACE

Q *Which is more beautiful from space, daytime Earth or night-time Earth? – Shreya, from Mellor Community Primary School in Leicester*

A The planet is stunning – both by day and by night. During my mission some of the things I loved observing at night were thunderstorms and the aurora. During the winter months we were fortunate to witness many spectacular aurorae. This was due to an increase in solar activity, which causes charged particles from the Sun to penetrate Earth's magnetic field and collide with atoms and molecules in our atmosphere. The result was a magnificent display of eerie green-and-red streaks of light that would snake beneath the space station or dance on the horizon.

At night, thunderstorms viewed from space were particularly impressive, too. On Earth, we only ever get to witness storms within our immediate local area of perhaps 50 or 60 km. From space, you can see entire storm fronts, stretching hundreds of kilometres. What's remarkable is the sheer quantity of lightning strikes that are occurring at any one time within a storm system. I remember seeing one storm front that stretched for several hundred kilometres along the coast of South Africa. The lightning was so intense that it resembled a strobe light continuously illuminating the night sky.

Also at night-time we get to see cities illuminated and signs of human habitation. Whilst this may look very beautiful from space, it's a reminder of just how much light pollution is caused by some of the vast urban areas of the world. By daytime it is much harder to pick out signs of human habitation. Instead we see the vast geological features of Earth, spanning entire continents and sculpted by four and a half billion years of nature's slow but steady grind. Some of the least-known regions of our planet make for the most stunning vistas from space: the volcanoes of Kamchatka, the glaciers of Patagonia, the dunes of the Sahara and the remote mountains of Kazakhstan and China all spring to mind.

There is certainly no denying the beauty of planet Earth. If I had to choose, I would say that Earth is most stunning by day. It really is a blue jewel, an oasis of life that shines out in stark contrast against the black void of space. I can only imagine that for those who have ventured farther out into space, such as the Apollo astronauts who went to the Moon, that view of 'home' must have felt even more precious.

This chapter is devoted to our home planet and the unique perspective gained by viewing it from space. But it is also devoted to the other view from the Cupola window – the vast inky vacuum, and the shimmering panorama of stars and planets that may one day be our new home. In order to survive in both of these environments we must learn to respect and protect them. And to do that, we need to understand the remarkable science that underpins it all.

Q *Can you see Earth's atmosphere from the space station – and what's it like?*

A Yes, we can see Earth's atmosphere from space. However, what struck me when I first saw the atmosphere was not the same feeling of serenity, awe and wonder that I experienced when looking down on the planet for the first time. Instead I remember thinking, 'Is that it? You've got to be kidding me! Life on Earth owes its existence to that thin strip

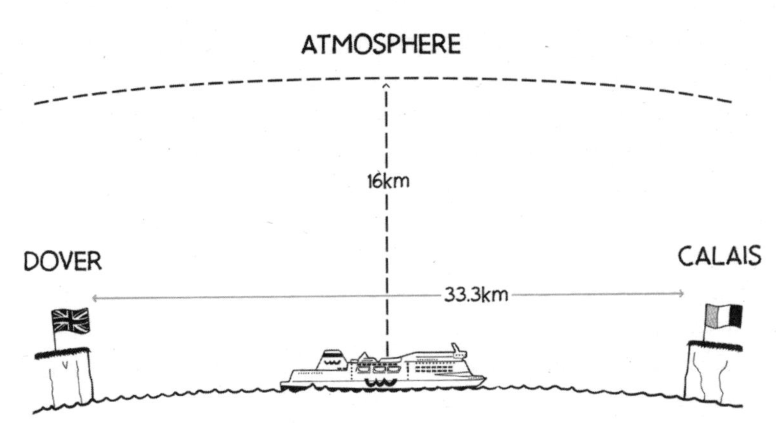

Thickness of atmosphere = ½ width of English Channel

of gas and . . . it's *tiny*.' Earth's atmosphere is really thin – if you imagine the planet as the size of a football, then the atmosphere is as thick as a sheet of paper. Most of the air is contained in a band only 16 km high. That's not even halfway across the English Channel from Dover to Calais!

To see the atmosphere in daytime, we have to look to the horizon where the curvature of Earth meets the blackness of space. Earth's atmosphere appears as a very thin band that looks white against Earth's surface, then gradually becomes light blue, dark blue and finally mixes with the black of space. The colours that we see are caused by the scattering of sunlight off the molecules of the atmosphere, called 'Rayleigh scattering'. If we look straight down or obliquely onto the planet, then we don't see the atmosphere – we just see the natural colours of Earth's surface. But we can also see clouds, weather systems, volcanic ash and sandstorms, which constantly remind us that there's a very active atmosphere down below. I was surprised one day, looking out over the Mediterranean, to see an enormous sandstorm stretching from the Sahara Desert across into southern France, Spain and Portugal. I watched as we moved around Earth and could see, when the sandstorm was on the horizon, how that part of the atmosphere appeared hazy with an orange hue, as the Sun reflected off the fine particles of sand.

At night we can only see the very top of the atmosphere. This appears as a thin greenish-orange strip of light and I was able to capture it in several photographs. This visual effect is called 'airglow' and is caused by a faint emission of light in the upper part of the atmosphere. Various processes create this emission, such as luminescence (caused by cosmic rays striking the atmosphere), chemiluminescence (caused by oxygen and nitrogen reacting with ions) and the recombination of atoms photoionised by the Sun during the day. Because this 'airglow' effect takes place in the upper part of atmosphere, when we see Earth's atmosphere at night it appears thicker than it does by day.

So, yes, we can see Earth's atmosphere – it is beautiful, but also incredibly thin and fragile; we'd do well to look after that precious band of life-supporting gas.

Q *Which destinations would you now like to visit for the first time on Earth, having seen them from up in space? – Anonymous*

A I've been fortunate to visit many beautiful places on Earth. I must be a sucker for punishment, as I tend to prefer cold, rugged and remote landscapes for my adventures. Some of my fondest travel memories are from a three-month expedition to Alaska at the age of 19, with a charity called Operation Raleigh (now Raleigh International). This is a sustainable development charity that works in remote rural areas to help communities manage natural resources, improve access to safe water and sanitation and protect vulnerable environments. In addition to their invaluable contribution to remote communities and the environment, Raleigh expeditions provide volunteers with amazing opportunities to build self-confidence and leadership through adventure and scientific exploration. Alaska certainly left a lasting impression on me, and during my time in space I always made a special effort when we passed the Aleutian Islands to grab a camera, head to the Cupola window and marvel at the striking beauty of Alaska's mountains, glaciers and rugged coastline. Each time was like taking a trip down memory lane.

Perhaps that teenage encounter with the great outdoors will go some way to explaining this list of places that I would now love to visit, having seen them from space (see photographs 31–34):

- The Andes, South America
- The volcanoes of the Kamchatka Peninsula, in the Russian Far East
- Nam Co Lake, China. In Mongolian it is known as *Tengur nuur*, meaning 'Heavenly Lake'
- Coast Mountains, British Columbia
- Lake Alakol and the Almaty province, Kazakhstan

Q *Can you see aircraft or ships from space?*

A It's not easy to see small objects with the naked eye from space. An exceptionally good human eye has a visual acuity of around one arc minute (or 1/60th of a degree). By doing some calculations, you can determine that at a distance of 400 km, the smallest resolvable size for the human eye (this is like the smallest pixel size of a screen) is 116 metres. This means that for the shape of an object on Earth to be discernible from the ISS, it would have to be bigger than 116 metres. But the real answer is more complex – just because you cannot determine something's shape doesn't mean you can't see it. The brightness of an object will also determine whether or not it can be seen. For example, you can look out on any clear night and see small satellites, no more than 10 metres in size, passing overhead in orbits more than 1,000 km high. This is because they are shiny and reflect sunlight at an observer.

In order to see a large container ship or even have a chance of spotting an aircraft from space, you need to know exactly where to look. One way to do this is to first pick out a ship's wake or aircraft contrails and then follow these telltale signs back to their origin. Then, if your eyes are better than mine and are feeling particularly fresh and sharp, you might be able to make out the tiny speck of a ship or an aircraft! At night, ships will sometimes stand out as tiny, single sources of light in an otherwise black ocean – or, in the case of fishing boats in the Gulf of Thailand, they can light up vast stretches of water with their green spotlights directed into the sea. From space, this looks as if some alien life-form is emerging from the depths, as the fishermen try to lure phytoplankton, which in turn attracts squid.

Much easier, of course, is spotting aircraft or ships using one of the many telephoto camera lenses that we have on board the ISS. These lenses have varying focal lengths to choose from, which give a magnified image. Anything larger than a 400 mm lens will enable you to see aircraft or ships in photographs. I took a picture of the port of Antwerp with a 500 mm lens, and not only can you see container ships clearly, but (unbeknown to me at the time) there was an aircraft passing overhead and you can clearly make out the contrail, followed by the tiny white outline of an aircraft. We also have stabilised binoculars on the space station, which are very useful for viewing approaching spacecraft, but there are no telescopes on board.

Q *In your photos of aurorae, is this how they appear to the naked eye or are the colours more intense because of the camera exposure?*

A For photographing the aurora, I found that a 0.5-second exposure and a light sensitivity setting (ISO) of 6400 gave a very close image to what we were seeing with the naked eye, in terms of colour and intensity. If anything (and I hate to disappoint you here) we see a more spectacular image with the naked eye. The camera doesn't do justice to the way the aurora eerily snakes and ripples, changing in both intensity and colour from the darkest part of the night orbit until daylight approaches.

Q *Can you see stars and planets from the space station, and do they look different?*

A Yes, astronauts can see stars and planets. In fact, from space we see them clearly as steady sources of light, as opposed to the 'twinkling' starlight that we often see on Earth. This is because turbulence in Earth's atmosphere causes most of that 'twinkling', as it refracts the light in different directions and makes the stars appear slightly less clear than when viewed from space. This is the reason why you find many of the world's observatories on mountain tops – in an attempt to reduce the amount of atmosphere through which the light has to travel. Another benefit of mountains is that there tends to be less light pollution.

Of course we have no atmosphere outside our windows in space, and it seemed to me that the planets appeared to be slightly brighter than when viewed from Earth – certainly Jupiter, Mars and Venus did. I was able to photograph Venus rising over Earth and also Jupiter, Mars and Saturn. Most of our windows look down on Earth – so although we see the planets rising and setting, it's much harder to see them when they're above the space station. You can see my photo of Venus rising just ahead of the Sun in the photo section (photo 30).

What's also interesting is trying to judge the distance of objects in space. Because there is almost no atmospheric interference, objects look sharper and clearer, even at a distance. When Cygnus, a commercial cargo spacecraft, departed from the space station after a resupply trip during my mission, we had the most spectacular view of it disappearing ahead of the space station. Of course it got smaller as it got further away, but it still looked incredibly sharp-focused and well defined, despite its increasing distance from us, which made it very hard to judge exactly how far away it really was.

Q *Why is it that in some pictures space looks black, with no sign of any stars or planets? – Gill Lee*

The reason you can't see stars in daytime photographs from space is that, when lit by the Sun, any foreground objects – such as Earth, the space station or my spacesuit – are many thousands of times brighter than the stars in the background. Earth is so bright that it swamps out most, if not all, of the light from stars and other planets. The stars don't show up because the camera cannot gather enough of their light in the short exposure times that we use for daytime photographs.

Our eyes work in a similar manner. The iris adjusts the central opening of our eye (the pupil), the diameter of which determines how much light strikes the retina. In bright daylight, our pupils contract in an effort to limit the amount of light entering the eye. In space, there is no chance during daylight of our eye being able to distinguish the paltry light from distant stars against the bright glare of the Sun. This is no different from being on Earth, where we would not expect to see stars in the daytime. It's just that in space the sky is black during the day, and this looks unusual both in photographs and to our eyes, because we are used to seeing stars against a black sky.

At night, our pupils dilate, allowing more light to enter the eye, in addition to striking more of the light-sensitive rods on the retina. This enables us to see objects that are less bright than the Sun, such as stars. You can test this concept for yourself by observing how many stars you can see on a bright night when there is a full Moon, compared to a much darker night with no Moon.

To take pictures of stars and planets from space, we need to wait until we are in Earth's shadow and then use longer exposure times (around 1–2 seconds) in order for the camera sensor to capture sufficient starlight. Because of the longer exposure times, we need to hold the camera extremely steady or the image will be blurred. Often for night photography I would use a stabilising 'Bogen arm' – a camera mount with a friction knob that allowed me to secure the camera at the desired angle and hold it much steadier than a human hand could manage. Some of my favourite photographs from space were taken using this method: pictures of the Milky Way rising over the horizon, or time-lapse sequences of the aurora, thunderstorms and Earth by night.

Q *Did being in space, and seeing Earth from space, change your perspective on the planet and life, or do you still feel the same?*

That's a great question and one that I get asked quite a lot. In some ways coming back down to Earth from space was, for me, a bit like visiting my old primary school. As a child your world is quite insular, usually revolving around home, school, family and friends. Your experiences at school are a huge part of those early formative years, but as you grow up and slowly become exposed to life outside that small bubble, your perspective changes. There's nothing like a visit to your old primary school to bring those early memories flooding back and to make you realise just how different your perspective of the world is today than it was then.

Going into space certainly broadens your horizons . . . quite literally! You get a more holistic appreciation of Earth and begin to feel strangely familiar with the planet. It may sound odd, but there are so many places that I now feel I know very well, despite never having set foot in those countries. Part of our morning routine on the ISS was to check the daily orbits to see which parts of the planet we might like to photograph that day: the Himalayas, the Bahamas, Africa, Alaska, Indonesia. As I reel off those names now I can recall the features of each location (and so many more) with astonishing clarity. I can visualise their valleys and glaciers, volcanoes and islands, mountains and rivers – all firmly etched in my memory.

When I first arrived on the space station our commander, Scott Kelly, had already been there for nine months. It was his second long-duration mission and his fourth spaceflight. I thought I was doing well when I could look out of the window and identify most of the major countries of the world. Scott floated past the window one day and casually noted, 'Ah, there's that nice beach on the coast of Somalia.' I'm not sure I ever got to that level of familiarity with the planet, but after six months there were not many places that I didn't recognise.

On the one hand, this fresh comprehension of the planet might be attributed to receiving a fairly extreme tutorial in geography. However, my experience involved far more than simply acquiring a talent for identifying locations on Earth. Seeing Earth from space gives you a sense of awareness and understanding as to our place within the solar system, the Milky Way and even the universe. Many astronauts have previously reported the same phenomenon and it has even been given a term: the 'Overview Effect' – a cognitive shift in awareness while viewing Earth from orbit or the lunar surface. I wouldn't dare to compare my experience of viewing Earth from 400 km with those of the Apollo astronauts, who travelled nearly 400,000 km away from Earth, to a point where the planet appeared as a minuscule disc occupying just a small fraction of a spacecraft window. But I think that both time and distance away from Earth contribute to this 'Overview Effect' and I certainly gained a new perspective of, and appreciation for, our small and fragile home during my time in space.

Perhaps Monty Python's 'Galaxy Song' sums up this feeling better than I have been able to articulate here – if you haven't heard it, then it's certainly well worth a listen, to add a bit of perspective to life!

Q *Does space smell?*

A This is a favourite question of mine and yet one of the hardest to answer. That's because *yes*, space does smell . . . but exactly *what* it smells of is much harder to put your nose on.

I smelt space on a number of occasions. The first time was after just

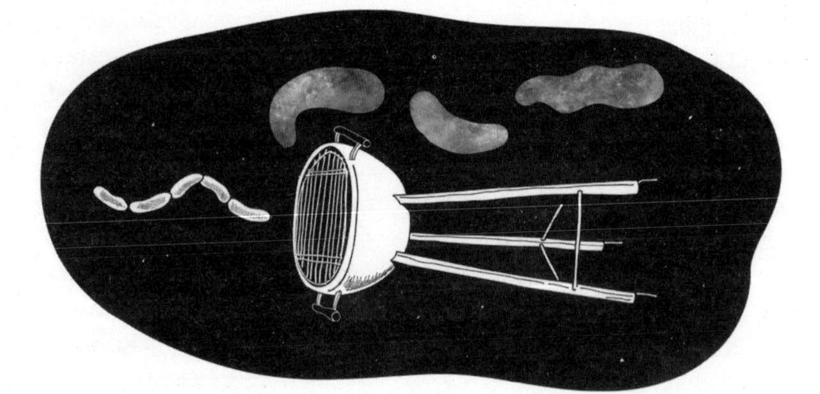

a few days on the Space Station, when I was helping astronauts Tim Kopra and Scott Kelly back in after their spacewalk. Subsequently there was a strong and distinctive smell whenever we opened the airlock after it had been exposed to the vacuum of space. I noticed the same odour each time I used the Japanese airlock, when transferring small satellites through it for launch or recovering experiments that had been outside the space station for several months.

The mystery scent is the topic of much light-hearted debate among astronauts. It has been described as seared steak, hot metal, welding fumes and barbecues, to name but a few. There are some suggestions that the smell may originate from the spacesuit itself, with certain components 'off-gassing', having been exposed to vacuum and thermal extremes. However, I smelt the exact same smell a couple of times inside an empty Japanese airlock following re-pressurisation. In my opinion, the smell of space is like static electricity. For example, when you take off a shirt or jumper and sometimes get a large static discharge – it has that kind of burnt metallic smell.

Actually what you're most likely smelling with static electricity is ozone. Ozone can occur naturally when high-energy ultraviolet rays (from the Sun, lightning or static electricity) strike oxygen molecules, splitting the molecule into two single oxygen atoms. A freed oxygen atom then combines with another oxygen molecule to form O_3 – ozone. Although ozone is present in the lower part of the stratosphere at around 20–30 km above Earth, it is not present at 400 km, so why would we smell it in space? Well, atomic oxygen *is* present in space. In fact, between 160 and 560 km, what little atmosphere there is consists of about 90 per cent atomic oxygen. It's possible that atomic oxygen is being introduced to the airlock when it is exposed to space, and on re-pressurisation it's reacting with oxygen molecules from the space station atmosphere, thereby creating ozone.

Perhaps the most wistful theory is that the smell of space is the leftover aroma of dying stars. There's an awful lot of combustion going on in the universe. Stars mostly comprise hydrogen and helium gas, powered by a nuclear-fusion reaction that can last for billions of years.

At the end of its life, as the hydrogen fuel is used up, a star will collapse on itself and undergo a violent supernova explosion, during which heavier elements such as oxygen, carbon, gold and uranium will be produced. All this rampant combustion produces smelly compounds called 'polycyclic aromatic hydrocarbons'. These molecules are thought to pervade the universe and float around for ever. So are we smelling the leftovers from some of the earliest stars, when we stick our nose into the airlock? Who knows.

Either way, I found it a rather pleasant smell and it reminded me a little of a British summer barbecue, burning sausages on a charcoal grill . . .

Q *Is it noisy in space?*

A Well, I'm not sure if this refers to 'out in space' or inside the space station, so I'll answer both. First, it's impossible for sound to travel through the vacuum of space. Sound waves need something to pass through, such as a solid, liquid or gas. Of course on Earth we are most used to hearing sound travelling through air. Sound is a vibration, where particles vibrate and collide with adjacent particles, propagating the sound as an audible mechanical wave. In the rarefied atmosphere of low Earth orbit, there are simply not enough particles to cause collisions and propagate the noise.

This is a pretty cool thing to witness in the vacuum of space. During a spacewalk, for example, I could knock my metal tether hook against a metal part of the space station and I would not hear a thing. The same metal-on-metal collision would cause a loud noise back on Earth. That is not to say that inside the spacesuit it is blissfully quiet. On the contrary, the spacesuit is working hard to keep you alive. This requires pumps, fans and airflow, all of which create quite a din. We wear a communications cap inside our helmet that incorporates a headset and microphone, in addition to providing some noise protection. So for astronauts working in the vacuum of space, things are not as quiet as you might imagine.

Things aren't much better inside the space station. We don't have to

wear communications caps, but the space station does have a plethora of ventilation fans, pumps and electrical equipment, all of which contribute to a fairly noisy environment. The background noise level varies around the space station, although I didn't really notice these variations much when floating between modules. The one exception was if someone was exercising on the treadmill. The treadmill generates a large amount of noise, as high as 85 decibels (recognised as the limit beyond which hearing protection should be worn), if someone is pushing themselves and running at high speed. To put that into perspective, most fighter pilots are exposed to about 80 decibels whilst wearing standard hearing protection inside a modern jet cockpit. To that end, astronauts exercising on the treadmill have specially moulded hearing protection that allows them to listen to some music or watch a movie/TV programme on the laptop, whilst at the same time protecting them from treadmill noise.

Generally, the rest of the space station is a more comfortable 50–60 decibels (comparable to a busy office environment) and our crew quarters have additional soundproofing in the walls and doors that reduce that further, to around 45–50 decibels.

Q *Is there gravity in space?*

A It's a common misconception that there is no gravity in space. In fact, gravity is everywhere! The great Sir Isaac Newton published his law of universal gravitation in 1687, supposedly after a close encounter with an apple. Newton described gravity as a force, stating that a particle attracts every other particle in the universe with a force that is directly proportional to the product of their masses, and inversely proportional to the square of the distance between them. This means that the force of attraction between two objects reduces (rather rapidly) the farther apart they are, but it never *completely* disappears. In this sense, gravity is the force that connects all matter in the universe.

Forces are nice, easy things to comprehend and we can understand the pull of the Sun that keeps the planets in their orbits, or the pull of Earth on the Moon. However, in 1916 another genius, Albert Einstein, complicated matters somewhat when he published his theory of general relativity. This had big implications for gravity. In essence, we now understand gravity not as a force, but rather as the curvature of space-time. Matter causes space-time to bend, warping the shape of the universe. Gravity is the effect that particles feel as they travel through this curved space-time on their journey through the universe. Newton's law still remains an excellent approximation of the effects of gravity in most cases, but when there is a need for extreme precision, or when dealing with very strong gravitational fields, then Einstein's relativity is required.

So you cannot travel through space without feeling the effects of gravity everywhere. On the International Space Station we are most definitely being affected by Earth's gravity, just as we are by the Sun's, by that of the other planets in the solar system and the supermassive black hole at the centre of the Milky Way (called Sagittarius A). Even as you read this book, the mass of your body is causing a small amount of curvature in space-time, which will be having an effect on the orbit of the ISS (admittedly, this is a rather small effect!).

The story of gravity is far from complete. Einstein's theory of general relativity has so far stood the test of time, and now scientists are in search of things like gravitational waves and gravitons, and are musing on the concept of gravity propagating through the universe at the speed of light. However, we still don't really know *what* gravity is; we only know how it behaves.

Q *Why do you appear 'weightless' on the International Space Station?*

A We appear weightless on the ISS because we are actually falling at the same rate as everything around us. Therefore if we were to try and stand on some scales in space, the scales would be falling with us and so they wouldn't register any weight at all – great news for anyone on a diet! By travelling very fast around Earth (about 27,600 km/h), we don't escape

Earth's gravity but instead, as we fall towards the planet, Earth curves away beneath us and we never get any closer to it. Our rate of falling exactly matches the curvature of Earth and we 'fall' the entire way around the planet. Since the space station and everything inside it is falling at approximately the same rate, we float and appear 'weightless'. We call this environment 'microgravity'.

So what would we really weigh on board the space station, if we weren't in constant freefall around the planet? Well, let's imagine we could build a tower 400 km high (the same altitude as the ISS) and weigh ourselves. Interestingly, we would still weigh 89 per cent of our weight on Earth – that's certainly not weightless! This is because 400 km is really pretty close to our planet, and the gravitational acceleration we feel from Earth is still 89 per cent of what it would be if we were standing on its surface. We calculate our weight by multiplying mass by gravitational acceleration; and so since Earth's gravitational acceleration hasn't reduced that much at the top of our 400-km tower, our weight hasn't changed that much, either.

Now imagine that you had taken a lift to the top of that tower. If the cable snapped and it fell back to Earth, then (if we ignored air resistance) you would be in freefall inside the lift and would enjoy the same feeling of weightlessness that astronauts do on board the space station – that is, up till the point at which you smash into the ground!

Q *How do you weigh yourself in space? – Michael, aged 29*

A That is a very logical question to ask, Michael – how *do* you weigh yourself in weightlessness? Well, in space we can't measure our weight directly, because we are in freefall and so our weight is essentially zero. However, we can measure how much of us there is, that is . . . our mass. By measuring our mass, we can determine what our weight on Earth would be. To measure our mass, we used a Russian device called a Body Mass Measurement Device (BMMD), which is a bit like a pogo-stick with a compressed spring. Astronauts first curl their body around the BMMD, hold on tight and then release the spring, gently travelling up and down whilst the device measures the frequency of oscillation. By first calibrating the BMMD and by knowing the stiffness of the spring, the device can accurately determine an astronaut's mass. We would do this three times and then average the result, although often the readings were within 0.1 kg of each other, such was the accuracy of the BMMD. Astronauts would usually 'weigh' themselves once every month whilst in space, a procedure fondly known as 'riding the donkey'!

Q *When you were in space, was there a risk of the ISS being hit by a meteor or piece of space junk?*

Actually, the space station gets hit by very small particles of debris quite frequently. Space debris encompasses both natural debris

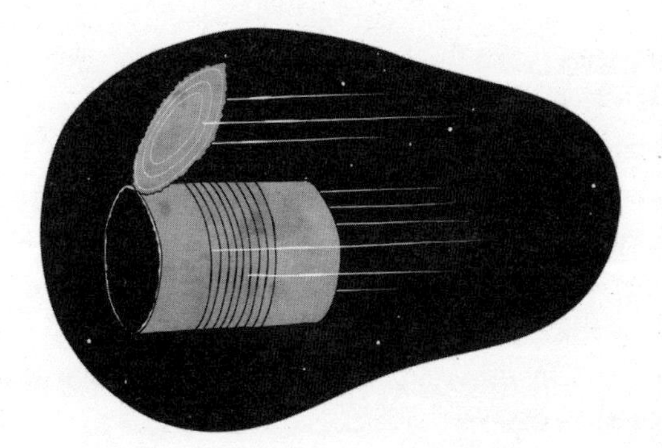

(micrometeoroids) and artificial (man-made) debris. Micrometeoroids orbit the Sun, whilst most artificial debris orbits Earth. Most of the time these impacts have no serious effect, and the space station is well protected by special shields that cover the pressurised modules that astronauts live and work in. However, there is a risk that something larger will strike the space station and cause damage. We see evidence of this on spacewalks, usually on handrails that have been struck by space debris and have left a small impact crater, often with sharp flayed metallic edges. Astronauts have to be particularly vigilant not to slide a glove over these razor-sharp protrusions and puncture their spacesuit. One of the Cupola windows has also suffered a debris strike, resulting in a small chip in the windowpane. But whilst no one likes to wake up and find a crack in a space-station window, it's not as bad as it looks. Each of the Cupola's seven windows has four panes of fused silica and borosilicate glass (made from silica and boron trioxide, which makes it particularly resistant to thermal shock), with a total glass thickness of more than 7 cm, and the chip has barely penetrated the first layer.

The problem is that when objects are travelling at hypervelocity, they don't need to be very large to cause significant damage. The chip in the Cupola was probably caused by a fleck of paint or a small metal fragment around a few thousandths of a millimetre across. So if something so tiny can cause damage to the space station, imagine what something 10 cm in diameter would do. Well, it would cause cataclysmic damage, penetrating straight through the space station with secondary effects that would shatter it into pieces.

The good news is that there are experts in Mission Control who warn us if there is a risk of collision – established as debris intruding into an imaginary pizza-box shaped 'keep-out zone' around the ISS (1.5 × 50 × 50 km). Some 23,000 pieces of space junk are being tracked by ground-based radar systems such as the US Space Surveillance Network and the European Space Agency's Space Debris Office in Darmstadt, Germany. When the risk of collision is high enough, the space station has to perform a 'Debris Avoidance Manoeuvre' (DAM), using thrusters from the Russian segment or a docked spacecraft to change the ISS orbit and avoid an impact. But an avoidance manoeuvre usually takes around 30 hours to plan and execute. If the debris has been spotted too late for the ISS to perform a DAM, the crew will be instructed to close all hatches between the various modules and shelter in their Soyuz spacecraft until the risk of collision has passed. The latest 'shelter-in-place' procedure was implemented in July 2015, when the crew received just 90 minutes' warning that a collision was possible.

The bad news is that there is a 'black zone' in terms of unknown risk of collision. Any object greater than 1 cm in diameter is large enough to cause catastrophic damage to the ISS and potential loss of life. This is the debris that the ISS is most at risk from, those pieces of 1–10 cm diameter that are hard to track, but highly likely to spoil your day. Using observations and computer models, it is estimated that there are 725,000 of these 'space bullets' between 1 and 10 cm in orbit around Earth. So, now that we know there is a risk of the space station being hit by a piece of space junk, the next question is highly relevant . . .

Q *What would happen if the space station was hit by space debris?*

A Let's imagine a larger object, say 2 cm in diameter, hits one of the space-station modules. Our first line of defence is the Micro-Meteoroid Orbital Debris (MMOD) shields. There are many hundreds of MMOD shields protecting parts of the space station and they differ in the materials used, mass, thickness and volume. Typical defences are Whipple and 'stuffed' Whipple shields. The basic principle of these shields is to have an aluminium 'bumper', which the debris strikes first. In addition to absorbing some of the impact, the bumper is designed to break the debris into smaller pieces, which then have less chance of penetrating the pressurised hull. Ideally you want as large a gap as possible between the bumper and the hull so that the broken-up fragments are spread over a wider area. The 'stuffed' Whipple adds some additional ceramic cloth and Kevlar fabric into this gap – materials often used in bulletproof clothing.

The European module, *Columbus*, is at the front of the space station and therefore at higher risk of a debris strike. However, even with shields of greater mass and a larger standoff distance, they will not stop a 2 cm-diameter object penetrating the hull. The first thing the crew will know about this is a big 'bang' from a strong acoustic shock wave as the hull perforates. If a crew member is unlucky enough to be in the same module, then they may witness an intense flash of light, before being struck by bits of the inside module wall breaking off (called 'spallation'), in addition to small fragments of the original debris. Some of these aluminium fragments burn quite actively, and the heat generated by the impact will also create a risk of fire. The perforation will typically be accompanied by rapid temperature changes and a decrease in air pressure, which can cause an internal fog. If the crew is unlucky, the perforation may be so large that a rapid crack growth occurs and the module 'unzips' or breaks apart completely. Such a dramatic break-up would probably be catastrophic for all crew members. However, assuming that the module maintains some sort of integrity, then the space station will begin to lose pressure, probably quite rapidly, and the crew would feel their ears 'pop' with the rapid change in pressure.

As a crew, we spend many hours training for a 'rapid depressurisation' emergency. In the scenario I've just described, it would be obvious to the crew which module had been struck and the most likely response would be for someone to shut the hatch immediately, sealing off that module and preventing the space station from going to vacuum. A similar situation occurred on 25 June 1997 on board the Mir space station. However, the rapid depressurisation was not the result of an impact with space debris, but of an impact with an approaching Progress resupply vehicle. Russian cosmonaut Vasily Tsibliyev had been instructed to remotely dock the cargo vehicle using a video image and a laser rangefinder. The problem was that none of the other crew members could see the Progress spacecraft through any of the windows, in order to use the rangefinder and report back its range. And without knowing how far away Progress was, it was impossible to calculate how fast it was approaching. The video image alone was useless in trying to judge the rate of closure and by the time Tsibliyev realised that the Progress was, in fact, approaching very rapidly, it was too late. Despite braking urgently, Progress collided with Mir with a great thump, rupturing the space station and knocking it into an uncontrolled spin.

Having seen the point of impact, Tsibliyev knew that it was the *Spektr* module that was leaking, but that module had not been readied for an easy hatch closure. It took the crew several minutes to cut the many cables that were strewn through the hatchway, until the module could finally be isolated to save the remainder of Mir. A lot was learnt from that near-disaster. Every hatch on the ISS is now designed to be closed in a matter of seconds. Normally, cables and other items are not permitted to be strewn through hatchways. Where this is unavoidable, a method of 'quick-disconnect' exists so that a hatch can be sealed with minimum delay.

The crew train for a rapid depressurisation in a very methodical fashion. First, having accounted for everyone and gathered in a safe haven, we work out how much time we have to deal with the problem until the pressure drop requires a complete evacuation of the space station. Then we check the integrity of the Soyuz spacecraft and try to establish which module is leaking. A slow leak may be hard to find, but by closing each hatch sequentially and monitoring to see if the pressure continues to drop or remains steady, it is possible to work our way through the entire space station until we find the leak. Of course we always work back towards our Soyuz spacecraft, being careful never to get on the wrong side of a closed hatch that would isolate us from our escape vehicle.

Q *How problematic is space debris? – Thomas Santini, from Robert Gordon's College in Aberdeen*

A Space debris is a huge problem. In addition to any naturally occurring micrometeoroids, for 60 years and more than 7,000 launches humans have been littering the orbital pathways around Earth with space junk – from rocket boosters and defunct satellites to tiny fragments and flecks of paint. There are estimated to be a whopping 150 million pieces of space debris larger than 1 mm, all trapped within a few thousand kilometres of Earth's surface. When launching satellites today, it is naïve to think that an attitude of 'big sky, small bullets' will prevent a significant impact from occurring – it's simply a matter of time. On 23 August 2016, operators at the European Space Agency's control centre in Darmstadt noticed that one of their Earth observation satellites, Sentinel 1A, had suddenly dropped electrical power and changed its orbit slightly. In just its third year of operation, this flagship satellite had been struck by debris and suffered damage to a 40 cm-wide area of the solar panel. It could have been much worse. The more we rely upon space-based assets in our day-to-day lives and for national security, the more serious the consequences of such collisions will be.

In 1978, NASA scientist Don Kessler recognised the danger posed by the high density of objects in low Earth orbit and the possibility of this creating a cascading chain reaction of collisions. Dubbed the 'Kessler Syndrome', the effect featured in the opening scenes of the 2013 film *Gravity*, as a cloud of debris destroyed a Space Shuttle and subsequently the ISS. Unfortunately the Kessler Syndrome is not just the stuff of science fiction. Half of all near-misses today are a result of debris from just two incidents. In 2007, China destroyed one of its own satellites with a ballistic missile. And in 2009, a US commercial communications satellite collided with a defunct Russian weather satellite. In spite of all this, the ISS is actually in an orbit where the amount of space debris is relatively low.

With the number of satellites in orbit looking to more than double in the next decade (to well over 18,000), this problem is only going to get worse. No single nation or entity is responsible for space. However, currently 85 countries are members of the Committee on the Peaceful Uses of Outer Space (COPUOS), set up in 1959 by the United Nations. Today various international organisations, space agencies and governments are working hard to understand the problem of space debris and make efforts to clean up space. The European Space Agency is at the forefront of developing and implementing debris-mitigation guidelines. Moreover, as part of their Clean Space initiative, ESA are planning the first-ever active debris-removal mission, e.Deorbit, with the goal of capturing a heavy ESA-owned item of debris and then destroying it in a controlled atmospheric re-entry. The US Defense Advanced Research Projects Agency (DARPA) is leading military efforts to find better ways of tracking space debris with a new ground-based 90-tonne Space Surveillance Telescope, which can track thousands of small targets and search an area larger than the size of the continental US in seconds. And since 2002, the Federal Communications Commission now requires all geostationary satellites to commit to moving to a graveyard orbit at the end of their operational life.

There is still an awful lot to do to reduce the threat posed by space debris, but doing nothing is no longer an option.

Quick-fire round:

Q *How many times did you go around Earth during your flight?*

A You can work this out for yourselves. First, you need to know how long I was in space – 186 days. Then you need to know that the ISS orbits Earth 16 times a day; or, more precisely, 15.54 times a day. So that's 186 x 15.54 = 2,890 orbits.

Q *How far did you travel during your time in space?*

A Well, distance is equal to velocity x time. We travel at 27,600 km/h. Then you need to work out how many hours there are in 186 days. The distance will be
186 x 24 x 27,600 = 123,206,400 km.

Q *Can you see the Great Wall of China from space?*

A Unfortunately not with the naked eye. I did try, but I couldn't make it out. However, once you know where to look, you can of course use an 800 mm focal-length camera lens and then take a photo of the Great Wall of China. Similarly, the Pyramids are impossible to spot with the naked eye. However, you can see the

Nile delta easily, which stands out against the desert like a sprig of broccoli and provides a great lead-in feature for knowing where to point the camera for that elusive Pyramid shot.

Q *Is there a formal protocol for 'first contact' with aliens?*

A This question made me laugh, although it's a great question to ask! The short answer is, disappointingly, 'No'. I would have loved to have sat through a briefing on what to do, should an alien life-form approach the space station but, alas, it was not forthcoming.

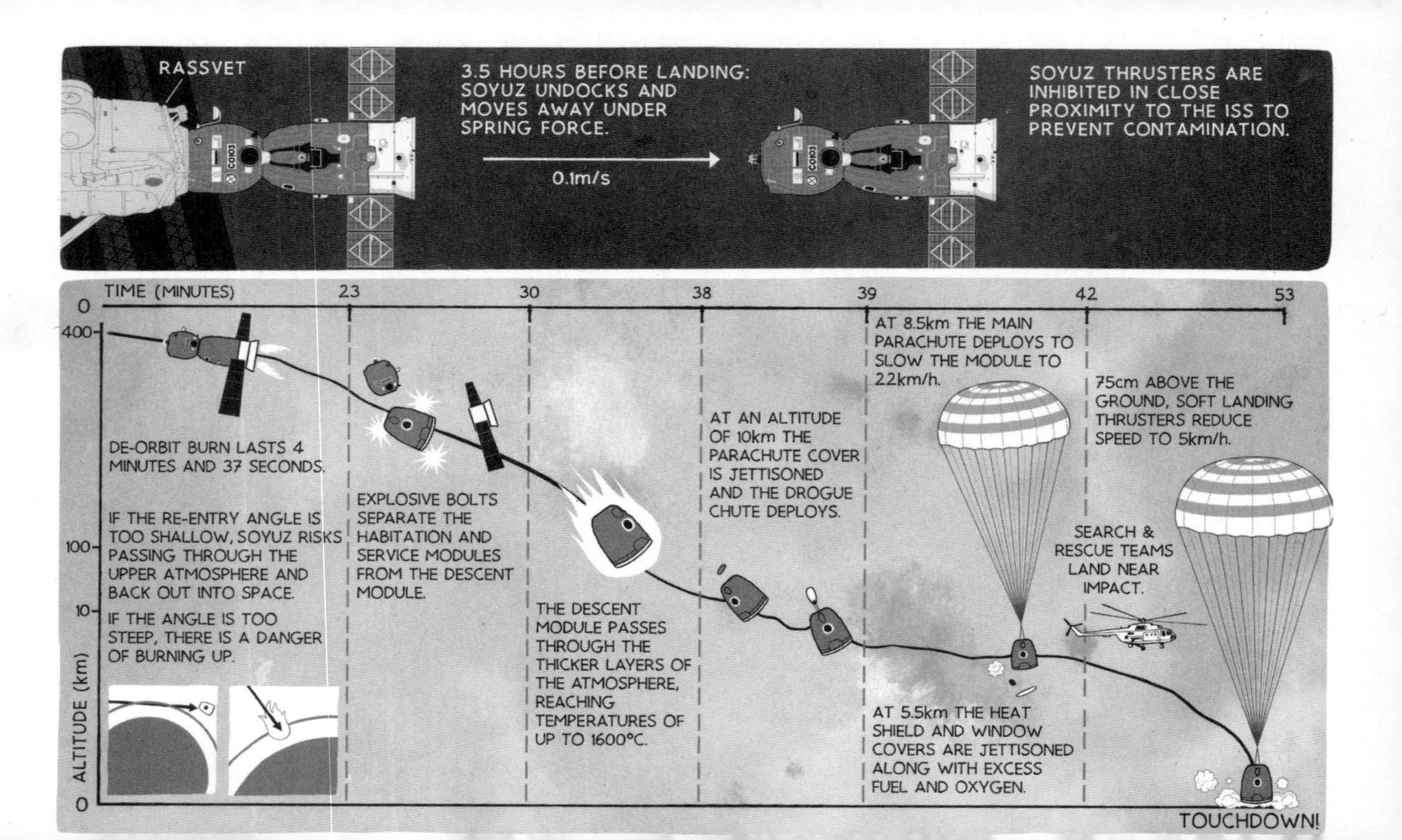

RASSVET
3.5 HOURS BEFORE LANDING: SOYUZ UNDOCKS AND MOVES AWAY UNDER SPRING FORCE.
0.1m/s
SOYUZ THRUSTERS ARE INHIBITED IN CLOSE PROXIMITY TO THE ISS TO PREVENT CONTAMINATION.
TIME (MINUTES)
0
23
30
38
39
42
53
400
100
10
0
ALTITUDE (km)
DE-ORBIT BURN LASTS 4 MINUTES AND 37 SECONDS.
IF THE RE-ENTRY ANGLE IS TOO SHALLOW, SOYUZ RISKS PASSING THROUGH THE UPPER ATMOSPHERE AND BACK OUT INTO SPACE.
IF THE ANGLE IS TOO STEEP, THERE IS A DANGER OF BURNING UP.
EXPLOSIVE BOLTS SEPARATE THE HABITATION AND SERVICE MODULES FROM THE DESCENT MODULE.
THE DESCENT MODULE PASSES THROUGH THE THICKER LAYERS OF THE ATMOSPHERE, REACHING TEMPERATURES OF UP TO 1600°C.
AT AN ALTITUDE OF 10km THE PARACHUTE COVER IS JETTISONED AND THE DROGUE CHUTE DEPLOYS.
AT 8.5km THE MAIN PARACHUTE DEPLOYS TO SLOW THE MODULE TO 22km/h.
AT 5.5km THE HEAT SHIELD AND WINDOW COVERS ARE JETTISONED ALONG WITH EXCESS FUEL AND OXYGEN.
75cm ABOVE THE GROUND, SOFT LANDING THRUSTERS REDUCE SPEED TO 5km/h.
SEARCH & RESCUE TEAMS LAND NEAR IMPACT.
TOUCHDOWN!

RETURN TO EARTH

Q *How long does it take to get back to Earth?*

A Returning to Earth from the ISS is a surprisingly quick journey. For Tim, Yuri and me, our six months of living on board the space station came to an end when our Soyuz spacecraft undocked at 05.46 GMT on 18 June 2016. We landed on the steppe of Kazakhstan, not far from where we had launched 186 days earlier, at 09.15 GMT the same day. It had taken just three and a half hours to return to Earth. That's shorter than a commercial flight from London to Moscow.

It felt surreal that less than 24 hours earlier I was having an ordinary morning, working in space, and now here I was, back on Earth. The day before, I'd had my hands full with a Japanese experiment that had been exposed outside the space station for several months. These fascinating studies are looking at how microorganisms such as fungi can survive in the harsh environment of space, in addition to capturing tiny organic-bearing micrometeorites in sticky 'aerogel', for assessing the possibility of interplanetary transport of life. Having retrieved the payload through the airlock in the Japanese module (*Kibo*), and whilst wearing full protective clothing, Jeff Williams and I were busy removing the precious scientific samples with the utmost care. I remember finishing that task around noon and thinking, 'Well, I'd better go and pack the last of my things and get ready to go home – I'm leaving in a few hours.'

That's one of the advantages of being able to return from low Earth orbit so quickly – in the event of an emergency, our Soyuz spacecraft becomes our lifeboat and we can be back on the planet within a matter of hours, not days. Sounds easy, right? If only that were the case. Like every part of a mission to space (and you might be detecting a theme here), the descent to Earth is extraordinarily complex for a spacecraft, meticulously planned and fraught with peril. It is also the most thrilling ride of your life. It certainly was for me. So, if you find yourself 400 km above Earth on the ISS, how do you get back? Well, in this chapter let's find out. Buckle up!

Q *Do you have to do any special training or preparation in space before coming home?*

A Yes. During the final two weeks of our mission there was time set aside in our work schedule for a number of tasks that helped us to prepare for the return to Earth. Most important was refreshing the skills needed to operate the Soyuz safely and to deal with any emergencies that might occur during our descent. It had been more than six months since we had last run through an emergency training session, back in the simulator at Star City in Russia. Whilst that training had drilled us until our response became second nature, after six months in space you begin to feel a little rusty. In addition to having some 'self-study' time to review the procedures and checklists for the descent, we also sat in our Soyuz as a crew and ran through a simulated undocking, descent and re-entry, whilst talking with our Russian instructor back at Moscow's Mission Control Centre. The Soyuz felt unusually cramped, having lived in the ISS for so long, and it was a stark reminder that soon we would be coming home.

There were a couple of things that we needed to check in the Soyuz. First, astronauts can grow up to 3 per cent taller during a six-month stay in space. Even for a vertically challenged person like me, at a mere

5 feet 8 inches, that could still mean a gain of 2 inches! This is simply because when the spine is offloaded in microgravity, the intervertebral discs 'relax', as do the tendons and ligaments surrounding the spine. This also happens on Earth, to a lesser extent, each time we sleep. On average we are 1 cm taller in the morning, but during the day gravity takes its toll and brings us down to size. When our Soyuz seat-liners are being moulded to our body before flight, the engineers take this spinal elongation into account and allow a couple of extra inches of headroom. Part of our preparation for returning to Earth is to check our seat-liners and ensure they still provide a good, secure fit that will protect us from the rollercoaster re-entry ride.

About three to four days prior to undocking, Yuri and Tim powered up the Soyuz spacecraft and checked out the motion control and propulsion system. Bearing in mind that the Soyuz had been in hibernation mode for six months, this was an important part of the preparation for return. When we undock from the space station, a mechanical spring force pushes the Soyuz away from the ISS at a gentle 0.1 metre/second. It would be a really bad day to discover only then that some systems were not functioning normally or that we were not able to control the spacecraft. Another piece of equipment that we have to check is the Sokol spacesuit. In the event of fire or a depressurisation of the descent module, this would save our lives and protect us against the vacuum of space. A week or so prior to returning to Earth, we donned our Sokol suits to check that they were still in good order and, most importantly, not leaking.

Yuri, as the Soyuz commander, was responsible for packing up the Soyuz spacecraft. Although there was not much room in the descent module, it was still necessary to bring back as much vital equipment and experimental data as possible. This included frozen samples of saliva, urine and blood that were needed urgently for life-science experiments. It was also possible to pack some rubbish into the habitation module, since the spacecraft is designed to separate into three modules prior to re-entry, with only the descent module making the journey back down to Earth. Any junk would simply burn up as it entered the atmosphere, along with the habitation module. However, the Soyuz had to be packed in accordance with a strict schedule approved by Mission Control. If the spacecraft's mass and centre of gravity were not correct, then the engine-burns that were calculated to bring the Soyuz back to Earth safely and to land on target would be wrong.

Q *Why don't you need a heat shield when leaving Earth, but you do need one on re-entry?*

A A heat shield protects a spacecraft when entering an atmosphere at extremely high speed (around 25 times the speed of sound, for a spacecraft returning to Earth from a low orbit, and even faster for vehicles coming back from the Moon or Mars). However, during launch we gain altitude at relatively low velocity, so the air does not get superheated. As a rocket climbs through the atmosphere it is subjected to a 'dynamic pressure' caused by the impact of air molecules against the nose-fairing. This will cause some aerodynamic heating due to skin friction, but nothing like the temperatures reached during re-entry. The Soyuz rocket encounters its maximum dynamic pressure (called 'max Q') about 49 seconds into launch and shortly after that it breaks the sound barrier. From this point onwards, although the rocket continues to accelerate, it is climbing higher and higher and therefore into less-dense air. Soon, just 2 minutes and 38 seconds into launch, the rocket is already about 80 km high and above most of Earth's atmosphere. With very few air molecules, there is almost no dynamic pressure and therefore no heating from friction. The nose-fairing is no longer required to protect the spacecraft and so it is jettisoned to save weight. That's the moment in the launch sequence when astronauts get their first look out of the Soyuz window and see space rapidly approaching.

By contrast, during re-entry there is little choice but to slam into the atmosphere at Mach 25 and use it as a brake to slow us down and return to Earth. To that end, the heat shield is designed to create drag to slow the spacecraft and dissipate the enormous heat generated by the shroud of incandescent plasma surrounding the spacecraft. Interestingly, in 1951 the National Advisory Committee for Aeronautics (which later became NASA) made the counter-intuitive discovery that a blunt shape made the most effective heat shield. Up until then, as aircraft were achieving faster and faster speeds, aerodynamic design had focused on low drag characteristics, such as sharp leading edges on aerofoils and pointed noses. However, when these models were tested in wind tunnels and subjected to supersonic speeds above about Mach 2.2, thermal heating became a real problem and some materials were melting. By using a blunt surface, the air can't get out of the way quickly enough. Instead, it acts like an air cushion and pushes the shock wave (and therefore the heated shock layer) forward. This prevents the hottest gases from coming in contact with the spacecraft. Instead, they move around the vehicle and dissipate into the atmosphere. This is one reason why the Space Shuttle did not have a sharp-pointed nose like a fighter jet – it would have melted during re-entry.

Q *Did you take any medicine to stop you feeling sick on the ride back to Earth?*

A Each astronaut will decide, in agreement with their flight surgeon, whether or not they wish to take medication to help with the feeling of nausea when returning to Earth. However, despite the re-entry being a fairly wild ride, I don't know of any astronaut who has actually felt sick during the descent or inside the Soyuz itself. Normally feelings of vertigo and nausea occur shortly after landing, as the body attempts to deal with a sudden change in the gravity vector. Astronauts on the ISS have access to standard medication used to treat motion sickness, such as Meclizine (branded as Bonamine, Dramamine, Sea Legs, etc.), or drugs containing Promethazine (such as Phenergran). It's important to know how you will react to any medication before you fly to space, and so during training we will try any medication that we think we may wish to use, to check our personal tolerance and for any side-effects. I knew that Meclizine worked well for me and, importantly, did not cause any drowsiness, so I took some prior to departing from the space station. I'm not sure it helped, though – I still felt pretty terrible about an hour after landing!

Another reason why astronauts returning to Earth feel unwell may be due to a lack of body fluid. There is little you can do during your time in space to prevent your body from losing fluid. It's an automatic reaction to microgravity that fluid gets redistributed more widely around the whole body, resulting in a decrease of plasma and blood volume of up to 20 per cent. Exercise will help to some degree to reduce this loss of plasma, but will not prevent it completely. This becomes a problem on landing, since a heart muscle that has already atrophied in space is suddenly subjected to an additional workload, having to pump less available blood to the head against the force of gravity. This can cause light-headedness, dizziness and fainting when standing upright, known as 'orthostatic intolerance'. To try and protect against this, astronauts take several salt tablets in the final hours of the mission and begin to drink about 2 litres of fluid (the quantities vary between astronauts, depending on their body mass). Fluid/salt loading prior to landing has proven somewhat effective in increasing blood-plasma volume, raising blood pressure and protecting against orthostatic intolerance.

Another way to protect against it is to wear antigravity suits or compression garments. Before returning to Earth in the Soyuz, the Russians provide a compression garment called the Kentavr. This is made of an elasticated fabric and consists of a pair of shorts that extend from the waist to the knee and a pair of gaiters that cover the calves. These can all be tightened using laces, to provide pressure on the body that resists venous pooling of the blood and helps to maintain arterial pressure.

Like most Soyuz crews, I wore the Kentavr under my Sokol spacesuit and took salt tablets prior to returning to Earth, to give my body the best possible chance to fight against those initial detrimental effects of gravity.

Q *How do you get back to Earth, and how fast were you going during re-entry?*

A The process of descending back to Earth starts at the same speed as the ISS, of course. However, to begin with it didn't feel very fast, since we were still up at 400 km and I had been used to watching the planet pass beneath me at that speed for such a long time. The initial engine-burn that began our descent was also not aggressive. We were strapped into our seats with the Soyuz spacecraft flying 'backwards', meaning that the main engine was pointing in the direction of travel. This is because in order to return from Earth's orbit, you need to *slow down* and let gravity bring you back home.

About 30 minutes prior to entering Earth's atmosphere the main engine ignited for a 4-minute and 37-second 'de-orbit' burn. This was enough thrust to slow us down by about 410 km/h. It felt like being pushed gently back into your seat and was nothing like the harsh acceleration of launch. The noise of the main engine lighting up was certainly very reassuring, as it had been idle for the past six months. If it had not worked, then the Soyuz did have smaller secondary thrusters that could have been used as a backup means of slowing us down. Following a successful 'de-orbit' burn, we were waiting for several things to happen. First, we were coming home – whether we liked it or not! By slowing down, the spacecraft was no longer in a circular orbit, but instead now followed a parabolic trajectory that would end in a collision with the planet; it could no longer escape the curvature of Earth. But the wacky thing about orbital dynamics is that as a spacecraft falls to Earth, it actually picks up speed – not enough speed to remain in orbit at a new lower altitude, but by the time our Soyuz entered Earth's atmosphere, it was travelling at about the same speed as the ISS again, despite the de-orbit burn that had slowed us down not long before. However, because we were now so much closer to Earth's surface (at around 100 km, as opposed to 400 km on the ISS) it felt *extremely* fast. Just prior to re-entry I looked out the window and was struck by an overwhelming thought, 'This is crazy, we're falling like a brick here' . . . which of course we were!

Before this happened, though, the spacecraft needed to separate into three parts. First, we had to depressurise the habitation module. In an emergency, this is not mandatory, but it would be a pretty explosive separation if the habitation module were fully pressurised! Separation occurred automatically 23 minutes after the de-orbit burn. This is when the fun started. Until this point the descent was really quite unremarkable. That changed the second the spacecraft blew itself apart. The Soyuz is held together by a number of bolts, with the descent module (where we were sitting) sandwiched between the habitation module and the service module. In order to separate, a number of pyrotechnic bolts and charges have to sever all the links between these three modules. I had been warned by previous crews that this is not a trivial event – and it didn't disappoint. A series of small explosions, which sounded alarmingly loud right next to my head, followed by a big jolt that shook the capsule, told me that our venerable spacecraft had been split into three parts. We were 139 km over the Arabian Peninsula and I remember seeing the Persian Gulf upside down as our severed descent module tumbled gently head-over-heels, waiting to slip into the clutches of the upper atmosphere.

Once in the grip of the atmosphere, as the service module and habitation module were being devoured by hot plasma, the descent module orientated itself so that the capsule was coming in heat shield first. Then it was simply a matter of letting the laws of physics do their thing – using the atmosphere as a natural air-brake to slow us down to around 800 km/h and welcome us back to Earth.

Q *How long does re-entry last and how many 'g's do you experience?*

A From atmospheric entry at 99.8 km until parachute opening at 10.8 km, the descent took 8 minutes and 17 seconds. However, we did not experience high g-forces for this entire time. After separation, the descent module tumbled as it plummeted Earthwards. Inside the Soyuz, I didn't notice this tumbling until I looked at the window. When I did, it was quite alarming to witness such 'out-of-control' attitudes, having been on a rock-steady space station for so long. After about four minutes of tumbling, the Soyuz reached the thicker part of the atmosphere, at around 80 km, and started to gain some aerodynamic control. As soon as the capsule oriented itself, heat shield first, we began to feel the onset of g-loading.

Thankfully, this commencement of g-loading was quite gentle, which gave us a few moments to get used to the feeling of weight again. Then, the 'g's began to build more rapidly. As this happened, we used the g-force to help push ourselves further into our seats and to pull the straps on our five-point harness as tight as possible. This was not to protect against the 'g's of atmospheric re-entry, since that's a fairly smooth ride. However, when the parachute opens (and for the actual landing itself) it is vital to be as secure as possible, to prevent injury. In the descent module, astronauts sit with their backs against the heat shield. This means that we feel the effects of deceleration as a heavy weight on our chests. Breathing became more difficult and it began to feel as if I was at the bottom of a rugby scrum, with several hairy forwards piled on top of me. Maybe those wet Saturday mornings playing rugby at school had been good astronaut training after all . . .

The g-loading peaked at a little over 4g, then slowly eased off. Because the g-force was through our chest, and not from head to toe, there was less chance of experiencing tunnel vision or 'blacking out', which can sometimes happen to pilots in high-performance aircraft. Despite feeling

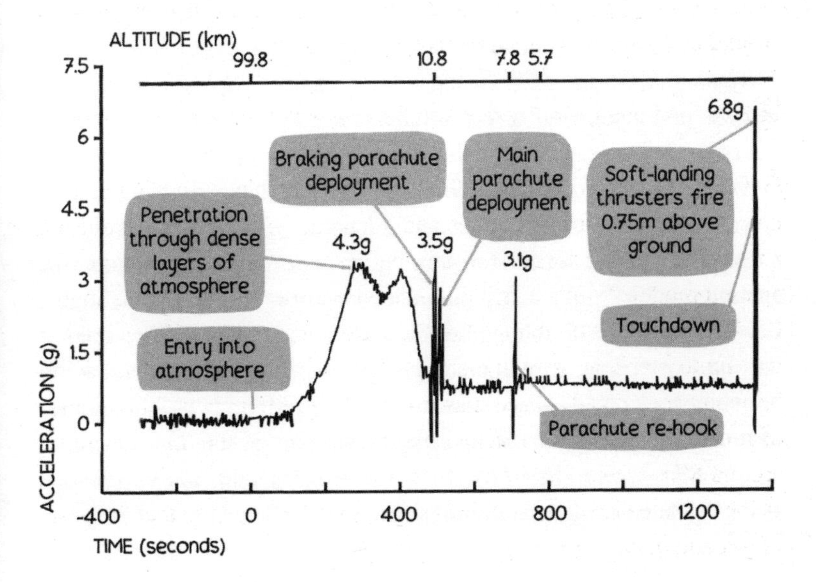

heavy and uncomfortable, the 'g's were easy enough to tolerate and our training in the centrifuge at Star City had prepared us well.

Q *How hot does the inside of the Soyuz descent capsule get during descent? And how is this controlled? – Dr Jaclyn Bell*

A The Soyuz has a thermal control system that operates on a similar principle to that of the ISS, albeit on a much smaller scale. Radiators located on the outside of the spacecraft (and therefore exposed to the cold vacuum of space) have fluid pumped through them. This cold fluid then flows inside the spacecraft and via a system of heat exchangers it mixes with the spacecraft's air supply. Cooled air is then circulated by fans around the modules and also fed into each cosmonaut's spacesuit for ventilation.

This is part of the 'active' cooling system. However, 'passive' thermal control is also used which consists mostly of multi-layer insulation (MLI) wrapped around the spacecraft. Much like the MLI used on our spacesuits, this material protects the spacecraft from both the extreme heat and cold of space. Under normal circumstances, the temperature inside the Soyuz is regulated between 18 to 25 degrees Celsius.

However, during descent and after separation occurs, the descent module no longer receives cooled fluid from the radiators, since this is all part of the service module which has now been jettisoned. Air continues to be circulated inside the descent module, but things start to warm up somewhat. Outside, temperatures can reach up to 1600 degrees Celsius as the heat-shield is exposed to an enormous amount of energy. Inside, I don't know exactly how hot it became, but the air circulating through my spacesuit felt very warm and I could feel myself perspiring. At the same time, I remember looking outside my window at the stunning firework display as pieces of MLI and pretty much anything else that could burn was being ripped from our capsule in a shower of sparks and flames. This lasted several minutes until about half way through the re-entry, when finally even the window covers themselves charred with the extreme heat and we could no longer see through them.

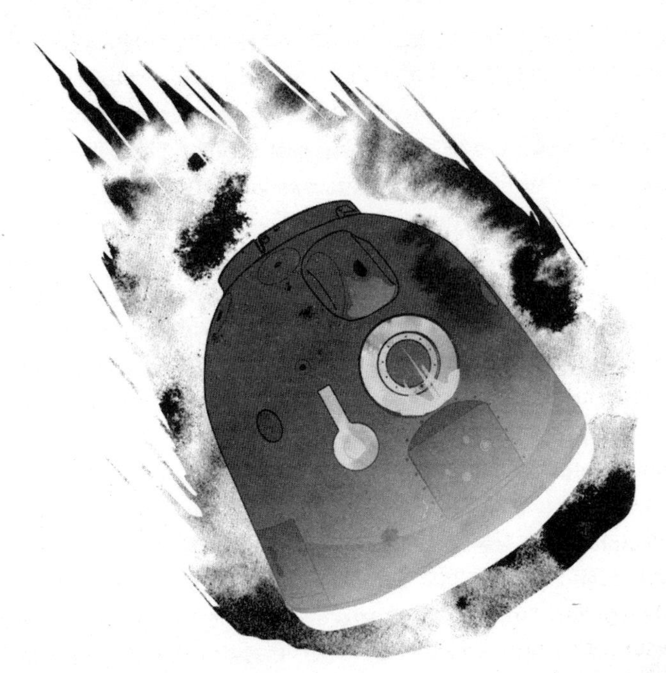

Did you know?

- During re-entry the plasma caused by shock heating of the atmosphere and ablation (the melting or wearing away) of the heat shield causes an interruption to radio and telemetry communications. This currently happens to all vehicles returning from space, although the duration varies, depending on the type of vehicle and re-entry profile. For our re-entry this communications 'blackout' lasted five minutes, during which Yuri continued to report g-loads, vehicle and crew condition whilst Mission Control waited for the spacecraft to emerge from the plasma phase and for communications to be restored.

Q *What did you enjoy the most: launch or re-entry? – Clare*

A Good question, Clare! I loved both the trip up and the trip back down, for different reasons. The launch was probably more exciting in terms of the feeling of raw power, acceleration and the anticipation of venturing into space for the first time. However, for simple rollercoaster kicks and thrills, then you should look no farther than being in a returning spacecraft when the parachute opens! Actually, it was not one parachute that opened, but a number of parachutes opening in a staggered release, designed to slow the capsule down from around 800 km/h to 324 km/h, when it becomes safe enough for the main canopy to deploy.

The atmosphere had done the bulk of the work in slowing us down, but at 11 km above Earth's surface we were still falling like a 3-tonne brick at just less than the speed of sound. The fun started when two drogue parachutes (or extraction parachutes) deployed. These drogue chutes pulled out a braking parachute, which sent the capsule into a violent spinning, rocking and shaking motion that lasted about 20 seconds. What made this wild ride even more pronounced and unpredictable was the fact that the capsule was not suspended directly beneath the parachute at that point, but was instead hanging about 30 degrees off-axis, until a later 're-hook' lined up everything for a safe landing. Simply put, it made for the ride to end all rollercoaster rides. I think NASA astronaut Doug Wheelock perfectly described what it felt like to be inside the capsule at that point when he said, 'It's like going over Niagara Falls in a barrel – but the barrel is on fire.'

My crewmate Jeff Williams had warned me that many astronauts comment about these 20 seconds of craziness during the braking parachute phase, but that I should also be prepared for a big jolt when the main parachute finally got pulled from the capsule and deployed. I had been watching the clock during the drogue-chute deployments as best I could, trying to focus on the small stopwatch in front of me. The violent shaking suddenly stopped and I could see that we had gone well beyond the 20 seconds of planned braking chute. It sounded as if air was still rushing past the capsule at high speed, but there had been no big jolt. Uncertain as to whether or not the main parachute had actually opened, I cautiously looked over to Yuri who, cool as ever, gave me a small nod – we were safe. So, yes: for kicks and grins, I would definitely pick re-entry over launch.

Q *The landing looks hard – did you have any bad injuries?*

A Thankfully not! Having said that, Tim Kopra joked that after landing he spent the first ten seconds patting his body down, doing a personal inventory to check that everything was in one piece. There's nothing subtle about a Soyuz landing – it's the same impact you would get from a minor car crash, and it definitely knocks the breath out of you. However, there are a number of design features that aim to prevent astronauts from being injured during landing. The priority is to slow the capsule down as much as possible. Once the main parachute has deployed, the heat shield and charred outer window glass are jettisoned. This reduces capsule weight and slows the rate of descent to about 22 km/h. At the same time, our seats rise automatically into a 'cocked' position, which helps to withstand some shock g-loads on impact.

Our seat-liners, personally moulded to our body shape in a bath of gypsum so many months earlier, were now an integral part of protecting us from the onrushing Kazakh Steppe. In addition, there are fabric knee braces that we tighten to prevent our legs from flailing on impact and causing injury. As we descended under the main canopy for about 15 minutes prior to landing, we made sure that there were no loose articles in the capsule that could fly around when we struck the ground. The commander, Yuri, wore a wrist altimeter so that he could give us an approximate countdown during the final 100 metres or so to touchdown. The three of us adopted a 'brace' position just prior to impact, with both arms crossed over our chests, holding firmly on to our flight checklists. We had been briefed to push our necks well back into our seat-liners, ensure that our mouths were closed and not have our tongue between our teeth. The worst thing you could do at this stage is look out of the window, as your neck would be in completely the wrong position to absorb the shock of landing.

Finally, when our capsule's gamma-ray altimeter detected that we were a mere 0.75 metres above the ground, a signal was sent to fire the 'soft-landing thrusters'. These solid-fuel thrusters fire immediately prior to impact, slowing the capsule to around 5 km/h. This is the main cause of the big puff of dust that you see when a Soyuz lands, rather than the actual landing itself. Although there was only a fraction of a second between the soft-landing thrusters firing and the capsule hitting the ground, it gave us just enough warning. Whilst astronauts frequently joke about the description of 'soft-landing' thrusters being a complete misnomer and more than slightly misleading, there's no doubt that without them the crew would almost certainly sustain some injuries.

When a 3-tonne capsule hits the ground, it doesn't really bounce. Instead the descent module simply slammed into the dirt, leaving a small crater on the outside and a slightly winded crew on the inside. One of the first priorities for the commander is to activate a button that cuts one of the parachute risers. This prevents the parachute from inflating on the ground and dragging the capsule across the surface, potentially causing injury to the crew. But you don't want to lose the parachute altogether since, if you are off-course or following an aborted launch, you may need the parachute to make a shelter whilst waiting for rescue. During the landing of my ESA classmate Alex Gerst in November 2014, the crew had difficulty in releasing the parachute riser and the capsule was dragged along the ground for quite a while. In a later interview, NASA crewmate Reid Wiseman said that being bumped across the steppe was the most 'dynamic' event of the entire return from space!

We landed with about 11 knots of wind, which was enough lateral velocity to cause our capsule to roll over on impact. However, our parachute riser released normally and, to our delight, we were injury-free. We had touched down 148 km southeast of the town of Dzhezkazgan in Kazakhstan – nearly 500 km east of where our journey began in Baikonur. Our Soyuz had rolled in a way that meant I was high up in the capsule, with Tim and Yuri below me. As we waited for the search-and-rescue crews to arrive on the scene and open the hatch, it was all I could do for those ten minutes to try and stop flight documents, checklists and other pieces of equipment falling on top of my crewmates . . . Gravity sucks!

Did you know?

- A report in 2016* highlighted that 37.5 per cent of US crew members experienced an injury of some kind, due to Soyuz landings. These were mostly minor in nature, with all injuries resolved within three months of landing. So I guess you could say that landing in a Soyuz is certainly not a trivial event.

Q *What happens if something goes wrong during re-entry and you land off-course?*

Good question! There were several things that could have gone wrong during the journey home, which is what makes it a particularly demanding phase of flight. The first thing that has to go well is the de-orbit burn. It must be extremely precise – a few seconds too long or too short would have meant missing the landing zone by hundreds of kilometres. If you take this to extremes, a de-orbit burn that is too short could place you into a re-entry angle that is too shallow to return to Earth. Contrary to popular belief, you would not 'skip' off the atmosphere like a stone skipping off a pond. Instead, the spacecraft would just pass through the less-dense upper layers without slowing down sufficiently to bring you home. You would simply head back out into space on a slightly elliptical orbit and probably come back to Earth a couple of hours later, in a completely uncontrolled and

* 'One Third of US Astronauts Injured During Soyuz Landings' by Keith Cowing, nasawatch.com, 27 October 2016.

catastrophic manner. On the other hand, a long de-orbit burn would slow you down too much and place you in such a steep re-entry angle that you would risk extremely high deceleration forces and very rapid heating of the heat shield, both of which could be disastrous. So a spacecraft returning to Earth has to accurately hit its re-entry 'corridor' – that narrow region that will permit a safe return to Earth.

The Soyuz spacecraft does not separate into three modules until after the de-orbit burn. That way, if something goes wrong, then you still have an engine and enough fuel, oxygen, electrical power, food, water (and a loo) for a few days – hopefully enough time to come up with another plan. However, assuming all goes well with the de-orbit burn, then 'separation' is the next main event. In the past, this has not always gone well. In fact Yuri was the commander for Soyuz TMA-11 when it returned to Earth on 19 April 2008 and suffered a partial separation failure. One of the five pyrotechnic bolts that secured the service module to the descent module malfunctioned, causing the spacecraft to enter the atmosphere with the service module still attached. A similar situation had occurred with Soyuz 5 in 1969. In both cases, after some tumbling, the spacecraft sought the most aerodynamically stable position – hatch first. Clearly, the hatch is not designed to sustain the punishing heat of re-entry. As the sole occupant of Soyuz 5, Russian cosmonaut Boris Volynov found himself in a capsule filling with dangerous fumes and smoke as the gaskets sealing the hatch began to burn. Furthermore, by facing the wrong way, he was being pushed outwards into his harness as the g-loads increased, instead of being pushed back into his seat. Thankfully, in both Volynov's and Yuri's incidents, the extreme heat of re-entry and aerodynamic forces caused struts to fail and the service module to detach before the hatch burnt through. This allowed the spacecraft to flip round and orient, heat shield first, to bear the brunt of re-entry.

In both these cases (and on other occasions when something has gone wrong), the Soyuz enters what is called a 'ballistic' mode. A ballistic re-entry is one in which the spacecraft relies solely on atmospheric drag to slow the vehicle. Under normal circumstances, the Soyuz actually produces a small amount of lift force, thereby reducing the descent angle slightly and making for a more comfortable ride (that is, less 'g') back to Earth. The US Apollo command module used this same technique for re-entry. However, during a ballistic re-entry, the capsule will fall to Earth at a steeper angle, with a higher rate of descent and higher deceleration forces. NASA astronaut Peggy Whitson was Yuri's Flight Engineer during Soyuz TMA-11's ballistic re-entry and she noted a punishing 8.2g on the display. Whilst this is not dangerous, astronauts have to withstand these g-forces for several minutes during re-entry, far longer than a fighter pilot might experience during a high-g manoeuvre. This is why astronauts train in the centrifuge at Star City and experience 8g for at least 30 seconds – it enables you to practise good breathing technique and prepare for what a ballistic re-entry may feel like. During the first US manned spaceflight, Alan Shepard experienced a whopping 11.6g during his planned ballistic re-entry.

The other problem with a ballistic re-entry is that the spacecraft will be way off-course. Normally the Soyuz is capable of making a fairly accurate landing. For our return, we were just 8 km from the intended touchdown point – that's pretty sharp shooting, for a descent that started halfway round the world! The reason why the Soyuz can be so accurate is partly due to a precise de-orbit burn, and partly to the ability to produce a small amount of lift during re-entry. The clever part is that for the first part of the descent the spacecraft will roll to the left and for the latter part it will roll to the right, thereby making a kind of S-shaped flight path through the atmosphere. By adjusting this roll angle slightly, both the azimuth (the horizontal direction) and elevation can be controlled during the descent, making this a simple but effective method for hitting the intended landing zone. During a ballistic re-entry, no lift is produced and therefore there can be no control. Soyuz TMA-11 landed 475 km away from its intended location. Nevertheless, the Russians prepare for this eventuality and, prior to return, the 'ballistic landing site' is determined, in addition to several other emergency landing zones around the world that could be used. The spacecraft will automatically transmit a locating beacon that alerts the search-and-rescue crews, once it has emerged from the communications blackout during re-entry.

As with the launch, the Russian Federal Air Transport Agency was responsible for providing Search and Rescue services for our landing. At the prime landing site, eight Mi-8 helicopters were waiting, in addition to four all-terrain vehicles and two fixed-wing aircraft that were airborne. These vehicles contained medical teams, search-and-rescue teams, management and press personnel. Furthermore, two Mi-8 search-and-rescue helicopters were located at the ballistic landing site, with another two located halfway between the prime and ballistic landing zones. In fact, before they fly to space, the crew even pack a 'ballistic landing bag', containing a flight suit, change of clothing, sunglasses and a wash kit. Sunglasses are essential, not in an attempt to look cool, but because the eye is an incredibly sensitive organ (particularly susceptible to radiation exposure) and, after six months of exposure to mostly artificial lighting, it is good to protect your eyes from sunlight as much as possible. Each crew member's bag was on one of the helicopters at the ballistic landing site, in case we had landed off-course.

So, if something goes wrong during re-entry, there are plenty of contingency plans in place to enable a safe return to Earth. However, you may be left waiting a little while until someone finds you. If all else fails, there's GPS and a satellite phone inside the Soyuz, so you can call your mum and let her know you're safe, and where you are!

Q *What was it like getting your first smell of Earth after being in space?*

A I was really looking forward to that first smell of fresh air on returning to Earth. That's not to say the ISS didn't smell pleasant, but it didn't smell of anything much at all. To me, it had a vaguely clinical, metallic smell when I first went on board, which I became accustomed to very quickly. Everything on the ISS is designed to minimise smells and prevent materials from off-gassing, and so your olfactory glands don't get much of a workout up there. With the exception of a few rare cargo deliveries of fresh fruit, we had been completely devoid of Earthly smells.

A few years earlier, in 2011, I'd spent seven days living deep down a cave in Sardinia, as part of ESA's human behavioural training. One of my most powerful memories of that experience was right at the end, when we emerged into the bright sunshine on a warm Mediterranean afternoon. The sights and smells were overwhelming. It was as if someone had turned the TV up to full contrast – the deep-blue sky and lush green trees all seemed to possess vastly exaggerated colour, after seven days of seeing nothing but varying shades of brown in dim light. I could smell Earth – truly smell the dirt, moss and lichen lying near the cave entrance. I relished those few moments, thinking how remarkable the world must seem to animals that have more acutely developed senses than us poorly equipped humans.

So it was with some anticipation that I prepared to inhale my first lungful of fresh air on return to my home planet. As it turned out, I should have held my breath! When the hatch opened, it was not a gust of clean, fresh air filled with the sweet smell of Kazakh grassland that greeted me. Instead we were welcomed by a pungent, acrid, burning smell that permeated the capsule. A mixture of scorched capsule – still searing hot from its plummet through the atmosphere – and burnt grass that had caught fire from the heat of the Soyuz came wafting in, shortly followed by a big Russian, who greeted us with a huge grin.

However, after several minutes, first Yuri, then I and finally Tim were pulled clear of the capsule and carried a short distance to where some chairs were waiting. That was the moment I had been waiting for and, finally, I was rewarded with a gentle breeze filled with all those wonderful Earthly smells.

Q *What happens after you've landed?*

A Having been pulled from the Soyuz, and following a short interview with waiting press, we were carried in our chairs to a nearby medical tent for a quick check-up. I was soaked in sweat, having felt very hot during re-entry and then having sat in a spacesuit for 30 minutes in the nearly 30 degrees Celsius heat of midday June in Kazakhstan. I was also dehydrated, and for once I was grateful to have an intravenous line plugged into a vein so that I could receive 1.5 litres of fluid. Getting out of my hot spacesuit and into a flight suit felt wonderful. Soon afterwards we took a bumpy ride to the waiting Mi-8 helicopter, which transported us about 400 km northeast to Karaganda airfield.

Having strapped into our Soyuz earlier that morning at about 03.00 GMT, and with no sleep prior to that, I didn't need much encouragement to close my eyes and sleep during the ride to Karaganda. Even after just a short rest, I felt much stronger standing and walking when we got to Karaganda. My flight surgeon kept a good grip on my arm to guard against my wayward balance, and to steer us through the crowd of people who had gathered to meet us. But already my body was starting to adjust to the feeling of Earth's gravity again. I remember my head feeling unusually heavy, and being conscious of my neck muscles having to work much harder than normal.

From Karaganda, Yuri went his separate way back to Star City, whilst Tim and I boarded a NASA aircraft that would take us to Bodo, in Norway. Bodo was the first refuelling stop for the NASA aircraft, and the point at which Tim and I parted company. Tim continued on his way back to Houston, whilst I boarded another aircraft to take me back to the European Astronaut Centre (EAC) in Cologne, Germany. This is the home of the European Astronaut Corps and would be my base for the next 21 days of rehabilitation and post-mission activities. I felt slightly bereft, saying goodbye to Tim and Yuri. We had lived and worked together for so long, and in such close proximity, that to part so suddenly and put several thousand kilometres between us was a little jarring. However, my sense of loss was not so much for their company, since I knew I'd be seeing Tim and Yuri again on numerous occasions in the future. It was for the mission. Each of us was heading back into the bustle and hectic schedules of post-mission debriefs, medical data collection and media interviews – all vitally important parts of an astronaut's job, but nothing that could possibly compare to the challenge of living and working in space. Saying farewell to Tim in Bodo was when I felt the mission had come to an end. I smiled – it had been a truly epic adventure!

Q *When did you have your first cuppa after landing?*

A My ever-thoughtful wife had given my flight surgeon some Yorkshire teabags so that I could enjoy a proper cuppa during the flight from Kazakhstan to Norway. My first non-recycled space-station-urine drink in six months. I think I had about three cups – it tasted great!

Q *When did you get to see your family again?*

A If my short stay at Bodo in Norway was tinged with sadness at the ending of our mission, it did not last for long. The European Space Agency aircraft that had just arrived to take me back to Cologne had a very special passenger on board – my wife Rebecca. Needless to say, we had one or two things to chat about during the short flight to Cologne, where we landed at around 3 a.m. to a warm reception from friends and colleagues from the European Astronaut Centre. Amongst the small crowd gathered to greet me were my parents – I walked straight over and gave them both a big hug. Our two young boys were still sleeping in my temporary crew accommodation next to EAC, and the one thing all parents know is that you don't wake a child unless you absolutely have to! So I grabbed a couple of hours' sleep before being woken by both my boys jumping on the bed and prodding their father, just to check that he had returned from space in one piece. Less than 24 hours earlier I had been orbiting the planet in a tiny spacecraft, and now here I was, being woken up – like many fathers on a Sunday morning – by my kids bouncing on the bed. All it needed was a cup of tea and a Sunday newspaper to complete the picture. It all felt very surreal, but wonderfully normal at the same time.

Saying goodbye to my family in Baikonur had been the hardest thing I have ever had to do. Despite all the training, preparation, checks, procedures and inspections, every astronaut knows the risks involved in flying to space. By strapping into a rocket, you are voluntarily rolling the dice and there's a chance you will not be coming home. I try not to take anything in life for granted and, as I hugged my wife and children that morning, I had every reason to be the happiest person on (or off!) the planet.

Q *What was the first 'proper' food you ate once you came back to Earth? – Scarlett Chatwin, aged nine*

A The first proper meal I had was Sunday lunch, in my crew quarters at EAC. Until then, I had mostly snacked on the flights from Kazakhstan to Cologne. Whilst in space, during one of my ham radio calls to a school, one student had asked me what food I was missing the most. I had replied that I missed fresh fruit and salad, of course, but that I also missed fresh bread and pizza. The bread that we get in space is 'Extended Shelf Life' bread, designed to prevent the growth of microbes. This is bread that has been formulated to reduce the amount of free water in the dough, by using binders such as glycerin, and has then been sealed in oxygen-free packages. The other option we have is to use tortilla wraps in place of bread. Although both are perfectly adequate for a quick peanut-butter and jam sandwich, it's not the same as eating nice fresh bread – and that is something most astronauts really miss in space. Perhaps John Young knew this already when, on 23 March 1965, he smuggled a corned-beef sandwich on board his Gemini 3 capsule in the pocket of his spacesuit.

His crewmate, Gus Grissom, was delighted when Young produced the snack in orbit and offered him a bite. Mission Control were less than thrilled, claiming that the crumbs could have wreaked havoc with the spacecraft's electronics.

I was pretty delighted, too, when my crew support team produced fresh pizza, fruit and salad as my first 'proper' meal back on Earth. I love Hawaiian pizza (I know, it's like Marmite – you either love it or hate it!), and the taste of warm fresh bread with chunks of juicy, fresh pineapple was truly delicious. It was the best pizza I have ever tasted!

Q *What was it like to walk again, after being in weightlessness for so long? – Loella Harris*

A For me, walking really did not feel very good for the first 48 hours after landing. This had nothing to do with lack of strength, balance or 'orthostatic intolerance'. Instead it was due to feelings of dizziness, nausea and vertigo. During an interview the day after I landed I described the feeling as 'the world's worst hangover'. I still stand by that description. There really is nothing else that compares to that debilitating feeling as my vestibular system reacquainted itself with gravity, and my brain tried once again to reconcile the various inputs that had changed so dramatically in the space of a few hours. As I sat in a chair, trying to minimise my head movements, I longed to be back in the liberating environment of weightlessness where it was virtually impossible to make myself dizzy. However, I also knew that the fastest way to readapt to Earth's gravity was for my body to relearn an old skill, and that wouldn't happen sitting in a chair. I had to get up and get walking.

At first, everything felt very heavy and clumsy and I adopted the 'astronaut stance'. This is a bit like the way John Wayne would walk after a long day in the saddle – feet wide apart, for extra stability, and a bit of a waddle. Once the feelings of vertigo subsided, it was actually quite fun to try out some balancing experiments. Standing on one leg was a challenge; looking up at the ceiling frequently had me falling backwards; and if I turned my head sideways whilst walking, then I would often wander off the pavement – I made a mental note not to try that one near traffic. I'm sure once or twice my ESA colleagues thought I had been enjoying a post-mission tipple as they watched me gently bounce off a wall whilst walking down the corridor, but alcohol was the last thing on my mind until my balance returned to normal. Thankfully this only took a few days.

Q *How did it feel to have your first proper shower after being on the ISS? – Lily*

A The first shower after being on the ISS was a mixture of pleasure and pain. Once I arrived in the crew quarters at the European Astronaut Centre, I was desperate to take my first shower in six months. I loved that feeling of hot water streaming over me – it felt so good. But whenever I stood up, I had feelings of dizziness and vertigo, which became worse with the sensation of water running past my ears. So I didn't spend long during my first shower – just enough to remind me of what I had been missing!

Q *Did you bring back any souvenirs from space?*

A I love this question – it makes me think of opening up a small market stall on the ISS selling postcards, trinkets and souvenirs. The trouble with the space station is that most of the stuff that would make great souvenirs is usually quite important, and the space agencies would get really upset if astronauts started stripping the place. After all, it costs a lot of money to fly things to space and so, once up there, they probably ought to stay there. Having said that, I did get to keep a few items that are very special to me.

I was able to bring back my space cutlery, which is really cool as it is engraved with 'Shuttle'. I guess we're still using up old stock and haven't yet opened the box containing the cutlery engraved 'ISS'! I also have a crushed Russian coin that I had in my pocket. That may sound a bit odd, but another Russian superstition is that it is good luck to carry a coin crushed by the train that pulls your rocket to the launch pad. I had asked one of my Russian friends if he could lay a coin on the railway track the morning that our Soyuz rocket 'rolled out' to the launch pad, as we were confined to quarantine at the time and it is considered bad luck for the crew to attend the roll-out ceremony.

However, the most special souvenir I was able to bring back was the Union Flag that I wore on my spacesuit during my spacewalk. As the first Union Flag to be worn in the vacuum of space, it holds a special significance, and to me it represents a new chapter in the UK's long and distinguished history of exploration and scientific research. I was fortunate to see a magnificent exhibition from the Royal Archives and the Royal Collection a few years before my mission, displaying artefacts from British exploration throughout history. I could think of no better place for this well-travelled flag to reside, and since returning from space I have had the honour of presenting this Union Flag to Her Majesty Queen Elizabeth II.

Q *Have you, or any other astronaut that you know of, ever come back to Earth and let go of something out of habit, expecting it to sit there, floating? – Ayda and Paul McCarthy*

A You know, I've heard many stories about astronauts dropping things when they return from space. This never happened to me, but I can see how it could happen with a light object, because you get so used to letting go of things in space. French astronaut Michel Tognini told me how he repeatedly dropped his cutlery during the first few days back, as he was so used to simply letting it float at the dinner table. However, I found that with anything heavy it was almost the opposite feeling: I would hold things in a death-grip, as I couldn't believe how much everything weighed back on Earth. One of the first things that Tim Kopra and I had to do after landing was an iPad-based experiment designed to evaluate our fine motor skills. We were on the NASA aircraft, having just left Kazakhstan, when someone handed me the iPad. I took it at arm's length and instantly regretted it, as I nearly dropped it on the floor. I think the flight surgeon must have thought

I had suffered severe muscular atrophy, as I couldn't even lift an iPad! Of course I had held the same iPad every day for six months in weightlessness, and it took me completely by surprise how heavy that thing felt back on Earth.

Q *What are the long-term health effects of spaceflight?*

A This is an important question, and one that every astronaut will have considered at some point during his or her career. If spaceflight were a drug, then the list of possible side-effects might well put you off taking that road to the stars! So before you go, let's have a look at what you might expect from a six-month dose on the International Space Station.

Muscle deterioration:

Symptoms: Without feeling the effects of gravity, skeletal muscle is no longer required to maintain posture and will begin to atrophy. Additionally, by not putting weight on the lower back or leg muscles that we use for standing up, these will also begin to weaken and get smaller. Astronauts can lose up to 20 per cent of their muscle mass in just 5–11 days in space.

Treatment: Regular exercise and a good diet will help to prevent muscle deterioration. The space station has the ARED 'multi-gym' device that astronauts can train on every day . This is particularly good for exercising the 'prime mover' muscles (quadriceps, biceps, triceps, pectorals, etc.), but it can be harder to stimulate some of the smaller stabilising, postural muscles and to maintain our core strength. The treadmill and bike machine also help to maintain good cardiovascular fitness and prevent the heart muscle from atrophying. Despite quickly regaining body weight that I had lost during the early part of my mission, I was not back to normal. My muscle mass had been redistributed differently and, although I felt physically stronger when I landed than when I launched (not surprising, when working out for two hours every day), if I picked up a heavy suitcase I could tell that my core stability had deteriorated. It took about two months until my core strength felt completely normal again.

Bone deterioration:

Symptoms: Bone is normally produced in response to mechanical stress, and our bones maintain resistance to fracture by remodelling through a balance of bone resorption and formation. However, in microgravity the decreased load on the bone disrupts this balance, resulting in a loss of bone tissue of about 1.5 per cent per month on the ISS. By comparison, an elderly person loses the same amount in about a year. Particularly susceptible is the pelvic region and lower spine, where astronauts can return from space with symptoms of osteoporosis. Furthermore, as bone mineral density reduces and is re-absorbed by the body, elevated blood calcium levels increase the risk of calcification of soft tissue and kidney-stone formation. And it's not just bone mineral density that's important. As new bone tissue forms in microgravity, the architecture of the bone itself can become altered, increasing the risk of fracture on return to Earth.

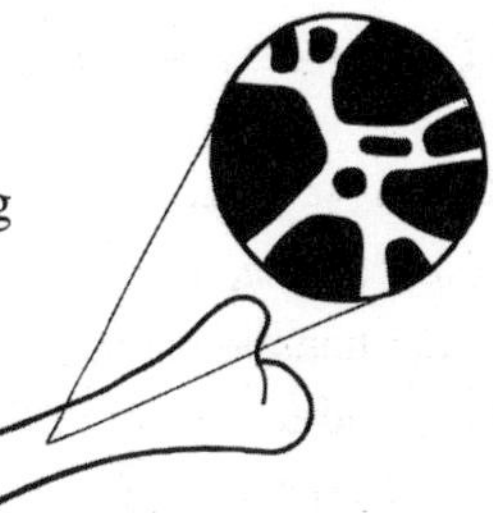

Treatment: Once again, exercise is our friend. By exercising our skeletal muscles and causing mechanical stress on our bones, we stimulate our 'osteoblasts' to build new bone tissue. As with our muscles, exercise can be more effective in some areas than in others, and exercise alone will not prevent bone deterioration. Astronauts usually take a daily vitamin-D supplement in addition to maintaining a good calcium intake in their diet, to help maintain healthy bones. A lower salt intake can help reduce bone loss, too – NASA reformulated more than 80 spacefood items to reduce the sodium content. Studies have also shown that taking bisphosphonate (a therapeutic agent used to treat osteoporosis patients) can help to prevent the loss of bone mass during spaceflight. I suffered the greatest bone loss in my femoral neck and lumbar spine. However, after just six months on Earth I had recovered 50 per cent of lost bone mass, and I'm on target for a full recovery within one to two years of landing, which is common in most astronauts. Much research has been conducted on the ISS investigating the loss of bone mineral density. These studies are helping not only to reduce bone loss in astronauts on long-duration missions, but also to advance the development of drugs to treat osteoporosis here on Earth.

Visual impairment:

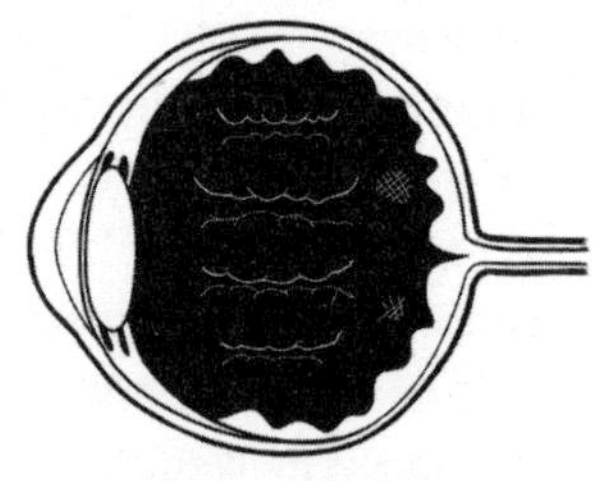

Symptoms: A more recent discovery is that spaceflight may impact upon vision. Changes ranging from disc oedema (swelling of the optic disc), posterior globe flattening (squashing of the back of the eyeball), choroidal folds (alternating light and dark bands in the retina) and cotton-wool spots (white patches on the retina) to nerve-fibre layer thickening and decreased near-vision have all been reported. In a survey of approximately 300 astronauts, 60 per cent of those who had flown long-duration missions experienced a degradation in visual acuity.

Treatment: It is still unclear which factors are causing these changes in vision. But it is likely that the microgravity-induced shift of body fluid, which affects cranial and ocular blood vessels, cerebrospinal fluid and causes elevated intracranial pressure is at least partly to blame. It has also been suggested that higher levels of atmospheric carbon dioxide, heavy resistive exercise or a high-sodium diet are contributory factors. Furthermore, individual susceptibility may be a factor, with some astronauts being genetically predisposed to developing visual impairment. Researchers are even looking into special clothing that astronauts could wear at night, which could be hooked up to a vacuum cleaner in an attempt to suck the blood and fluid towards the feet and ease the pressure on the heart and brain whilst you sleep in space!

Radiation exposure:

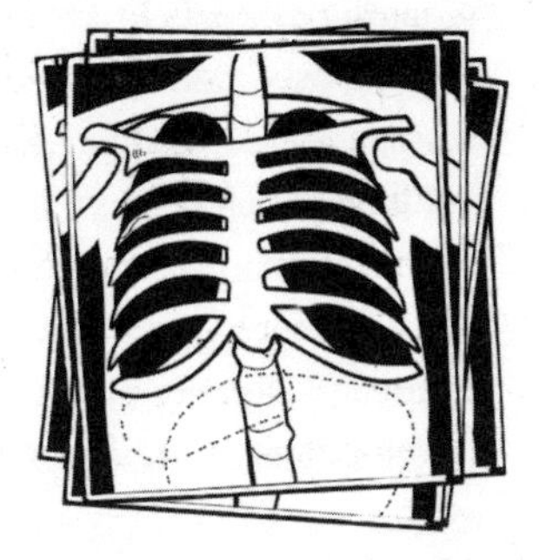

Symptoms: Thanks to Earth's magnetic field, we are largely protected from radiation coming from space whilst living on the planet. However, on the ISS astronauts are more susceptible to solar radiation and to radiation from galactic cosmic rays (high-energy particles coming from deep space). The fast, heavy ions that make up cosmic rays not only damage body tissue, but also cause trouble when they collide with the aluminium hull of the ISS, releasing a shower of secondary particles into our living quarters. On average, astronauts on the ISS are subjected to about 0.7–1 millisievert of radiation per day. A six-month stay on the ISS is equivalent to about 60 years of radiation that you would receive from natural sources on Earth. Or, to put it another way, being in low Earth orbit is like having eight chest X-rays every day.

Treatment: The best treatment for radiation is to limit exposure as much as possible. To that end, parts of the ISS have polyethylene shielding, which reduces the impact of the secondary neutrons emitted when it is struck by cosmic rays. The good news is that the radiation environment on the ISS is being closely monitored. Not only are there a plethora of radiation monitors in every module, but each astronaut carries a personal dosimeter at all times. Additionally, during a spacewalk each crew member carries a separate dosimeter, because we receive a higher dose when we leave the sanctuary of the pressurised modules. The bad news is that there's no way to sugar-coat the fact that astronauts do receive a large dose of radiation during long-duration space missions. It is hard to gain a consensus on how to convert that radiation dose into cancer risk, but NASA requires that the increased cancer risk for astronauts due to radiation exposure will not be greater than 3 per cent above the estimate for the general population.

Vascular ageing:

Symptoms: As humans get older, our arteries stiffen, which causes an increase in blood pressure and elevates the risk of cardiovascular disease. Recently it has been observed that astronauts returning from the ISS have much stiffer arteries than when they went into space. In terms of arterial stiffness, six months in space is the equivalent of 10–20 years of ageing on Earth!

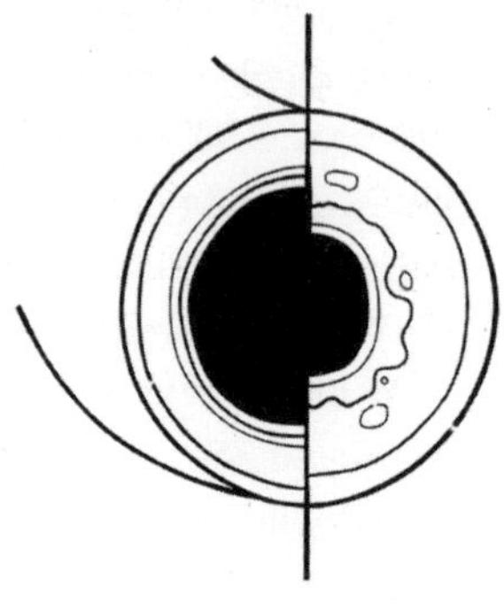

Treatment: The good news is that on return to Earth the ageing process begins to reverse, and astronauts can expect their arteries to recover to pre-launch condition within a few months. By observing these changes, scientists will be able to better understand the mechanisms that underpin arterial stiffening, in order to develop countermeasures to slow vascular ageing. I volunteered for a Canadian Space Agency experiment called 'Vascular Echo', which is leading research in this area, and I performed the first ultrasound session for this experiment whilst on board the ISS. In addition to understanding the risks to astronauts, the goal of this research is to slow vascular ageing and improve the health and quality of life for everyone on Earth.

Back/neck pain:

Symptoms: Going into space can be a real pain in the neck . . . literally. The impact of an elongated spine, weakened stabilising muscles (especially those supporting the spine) and altered body posture can play havoc during the first few weeks in space, but it can have even longer-lasting effects on return to Earth. More than half of astronauts report mission-related back or neck pain. During the first year after landing from a long-duration mission, astronauts are four times more likely than folk on Earth to suffer a herniated disc.

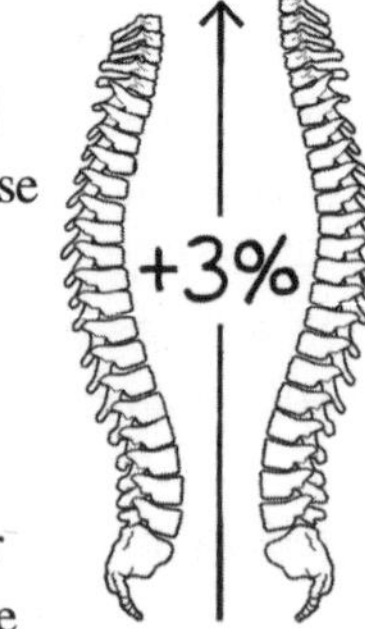

Treatment: In addition to the standard exercise programme in space, astronauts can use bungees, 'Therabands', stretching and even yoga exercises to try and stimulate the core muscles and alleviate back or neck pain in orbit. During the post-mission phase, exercise and physiotherapy specialists work with each astronaut to evaluate the degree of muscular atrophy and develop a tailored rehabilitation programme. But it can often take several months to recover fully from back or neck pain associated with spaceflight.

Depleted immune system:

Symptoms: Research has shown that our immune system can suffer for a number of reasons: stress, inadequate sleep, isolation, exposure to radiation or poor diet, to name a few. All of these conditions exist in space, in addition to microgravity, which places an unknown stress on the immune system. Sure enough, data confirms that astronauts' immune systems are getting confused in orbit. Some immune-cell function is lower than normal, whilst other cells have heightened activity. Clearly, a depressed immune system won't respond correctly to threats – placing an astronaut at risk of infection. However, heightened cell activity can lead to an excessive response, resulting in allergy symptoms and rashes, which have been reported by some crew members.

Treatment: Scientists are still investigating this, to try and better understand why the immune system becomes confused and how to protect astronauts on long-duration missions. Countermeasures could include improved radiation shielding, nutritional supplementation, pharmaceuticals, and more. Studies have also indicated that pre-exposure to radiofrequency radiation may trigger a response in our immune system, providing increased resistance to subsequent infection. These investigations may provide great benefits to Earth-based medicine, as we learn how and why changes to our immune system occur.

★

So, after reading this answer – hands up who wants to take a trip to Mars?

Did you know?

- Many of the negative health effects of spaceflight can be prevented to some degree by simulating gravity using a centrifuge in space. In Andy Weir's novel *The Martian*, which later became a blockbuster movie, the Earth–Mars transfer spacecraft incorporated a rotating module called *Hermes*, simulating about 0.4g – close to martian gravity. That's a really great idea, but a centrifuge section adds a lot of complexity and cost to spacecraft design.

AFTERWORD: LOOKING TO THE FUTURE

Q *If your next mission is not to the ISS, will you have to undergo a different type of training for wherever you may go to? – Mary Bainbridge*

A This seems like a great question to bring this book to a close – looking forward to the exciting future that lies ahead for human spaceflight and space exploration. The short answer to your question, Mary, is *yes*: for destinations other than the ISS, some of the training will be different. In fact even for a future mission to the ISS, the training could be significantly different. To give you a flavour of how and why this preparation will vary, here's a quick look at the assortment of spacecraft and space stations that astronauts may need to train on, in the very near future.

Commercial crew transportation

The United States is once again close to being able to deliver crews to the ISS on rockets launched from American soil. In September 2014, NASA selected two companies, Boeing and SpaceX, to provide a capability to launch and return four astronauts to the ISS. This will increase the total ISS crew to seven – adding 40 hours per week of crew time dedicated to scientific research. The two spacecraft – Boeing's CST-100 Starliner and SpaceX's Dragon – will end the current reliance on the Russian Soyuz rocket as the sole means of transporting crews to and from the ISS. It is expected that by 2019 both of these vehicles will be in operation, with crews already in training on these spacecraft.

Low Earth orbit

Operation of the ISS has been extended until 2024. As the benefits of microgravity research multiply and become more widespread, there is growing interest from the private sector. Two commercial companies, Bigelow Aerospace and Axiom Space, have plans to build and operate commercial space stations in low Earth orbit. Bigelow's BEAM module is already attached to the ISS for a two-year testing period, and Axiom Space plans to use the ISS as an initial hub for its first space-station module in the early 2020s. There are discussions about extending the ISS lifetime through 2028, to allow for a gradual transition to the private sector of microgravity research platforms in low Earth orbit. What is certain is that missions to the ISS in the 2020s will continue to be extremely busy, exciting and dynamic, as the commercial sector expands its foothold in space.

Lunar exploration

There is good reason why the national space agencies are supportive of transitioning to commercial space stations in low Earth orbit – it allows them to focus valuable and limited resources on the next stages of human exploration in our solar system. This comes in the shape of a new heavy-lift rocket: NASA's Space Launch System (SLS), which is larger and more powerful than the Saturn V that launched the Apollo missions. The first five SLS launches will see the assembly of a Deep Space Gateway (a small space station with power/propulsion, habitation, logistics and airlock modules) in orbit around the Moon. Beginning in 2019, these missions will not only pave the way for scientific research in deep space, but will also provide the opportunity for a return to exploring the lunar surface, and will act as a stepping stone to Mars. Following the first unmanned test flight, SLS will carry the Orion capsule and four crew on subsequent missions of up to six weeks' duration as the Deep Space Gateway is assembled, with a planned completion date of 2026. Astronauts will tend the new space station on missions of several weeks' duration, but will not reside there year-round. Although this concept is being driven by NASA, it requires strong partnership from other national space agencies, academia and the private sector. The European Space Agency already plays a vital part in these missions, providing the European Service Module for the Orion spacecraft.

Mars exploration

Ever since Neil Armstrong set foot on the Moon on 20 July 1969, people have looked to Mars as the next step in human exploration of our solar system. A roadmap is finally starting to emerge that will fulfil this long-held ambition. SLS will launch a Deep Space Transport spacecraft, to be assembled at the Deep Space Gateway, as early as 2027. Following the launch of logistics modules and a year-long crewed 'shakedown' mission in the Moon's vicinity, there are plans for this 41-tonne vehicle to carry a crew of four on a mission to Mars and back in 2033. Lasting for up to three years, this mission will not land, but instead will orbit the Red Planet and return to the Deep Space Gateway. The lessons learnt from this mission will pave the way for the ultimate goal of setting humans down on martian soil, as we take our first steps in colonising another planet.

And it's not just the national space agencies that are pushing forward with plans to explore further into our solar system and to set foot on Mars. Elon Musk (CEO of SpaceX) has made no secret of his ambitions to colonise Mars and ensure that humans become a multi-planetary species. This is far from idle talk. SpaceX have already built and are testing a methane-fuelled Raptor engine for Musk's 'Interplanetary Transport System'. This powerful engine will have more than three times the thrust of the current Merlin 1D engine that powers SpaceX's Falcon 9 rocket to the ISS. The reusable first-stage booster for this new rocket will have a staggering 42 Raptor engines, developing nearly four times the thrust of the Saturn V rocket that launched the Apollo missions to the Moon and providing a sustainable launch capability.

SpaceX isn't the only billionaire-backed company that is planning to reduce the cost of access to space and push the boundaries of human exploration. Jeff Bezos (founder of Amazon) has his own rocket company, Blue Origin, which is currently developing a series of new rockets, with its sights set on returning to the Moon and expanding the human presence in the solar system.

Furthermore, Sierra Nevada Corporation continues to develop its Dream Chaser spacecraft, and in 2016 it was awarded a NASA contract to provide a minimum of six commercial resupply missions to the ISS between 2019 and 2024. And with companies such as Virgin Galactic, Blue Origin and XCOR developing technologies that promise soon to offer a tantalising experience of space for hundreds of people, the next few years are about to get extremely interesting for human spaceflight.

The starting gun on this new 'space race' fired several years ago. It is a race that offers not only competition, greater sustainability and low-cost access to space, but also exciting opportunities for collaboration, new partnerships and international cooperation. The race is just starting to hot up, and a new dawn is approaching for space exploration. It is no longer a question of *if* we will colonise the Moon and Mars, but *when*.

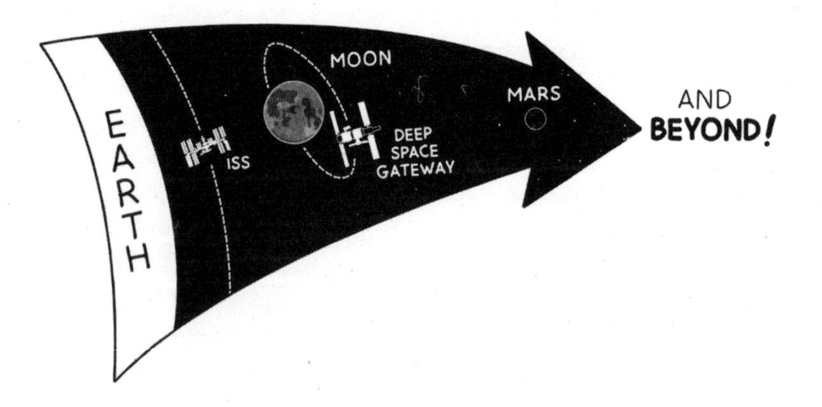

ACKNOWLEDGEMENTS

First and foremost, I'd like to thank everyone who has contributed to this book by showing a curiosity in human spaceflight and for asking a great variety of interesting, challenging and often humorous questions. I've enjoyed answering every one of them. For keeping a watchful eye on my answers I turned to two of the most intelligent people I know – thanks to my father, Nigel Peake, and best man, Dr Ian Curry, for your words of advice. Thanks to Carl Walker and Rosita Suenson of the European Space Agency for fact-checking and helping to launch this project. For the brilliant illustrations in this book, I'd like to thank Ed Grace, and Joe Kirton for helping out.

This book would not have been possible without the help of the dedicated team at Cornerstone, Penguin Random House. Thanks to my editor, Ben Brusey, who now knows enough about space to qualify as an astronaut. To Jason Smith for the cover, Mandy Greenfield, Joanna Taylor and Katie Loughnane for copy-editing, Becky Millar for organising the picture sections, and Linda Hodgson for producing the book. I'm grateful to Charlotte Bush in Publicity, Rebecca Ikin and Hattie Adam-Smith in Marketing, Aslan Byrne and the Sales team, Pippa Wright in Rights, and Susan Sandon for her support. I'm grateful, too, to the team at Little, Brown in the US.

I'd also like to thank teachers, instructors and mentors from all walks of life who devote so much of themselves to helping others reach their goals. Your patience and enthusiasm when answering our questions continue to guide and inspire us.

Finally, I'd like to thank my wife, Rebecca, for her endless support and encouragement during what has been a much bigger undertaking than I'd anticipated!

PHOTOGRAPHY CREDITS

1. © ESA – S. Corvaja
2. © NASA
3. © NASA – L. Harnett
4. © Getty
5. © ESA
6. © NASA
7. © GCTC
8. © ESA
9. © UKSA
10. © GCTC – Yuri Kargapolov
11. © GCTC – Yuri Kargapolov
12. © NASA – B. Stafford
13. © NASA
14. © GCTC
15. © UKSA – Max Alexander
16. © NASA – Victor Zelentsov

17. © ESA – S. Corvaja
18. © ESA – S. Corvaja
19. © ESA – S. Corvaja
20. © ESA – S. Corvaja
21. © ESA – S. Corvaja
22. © Getty
23. © ESA / NASA (picture taken by ESA Astronaut Tim Peake)
24. © ESA / NASA (picture taken by NASA Astronaut Tim Kopra)
25. © ESA / NASA (picture taken by NASA Astronaut Tim Kopra)
26. © ESA / NASA (picture taken by ESA Astronaut Tim Peake)
27. © ESA / NASA (picture taken by ESA Astronaut Tim Peake)
28. © ESA / NASA (picture taken by ESA Astronaut Tim Peake)
29. © ESA / NASA (picture taken by ESA Astronaut Tim Peake)
30. © ESA / NASA (picture taken by ESA Astronaut Tim Peake)
31. © ESA / NASA (picture taken by ESA Astronaut Tim Peake)
32. © ESA / NASA (picture taken by ESA Astronaut Tim Peake)
33. © ESA / NASA (picture taken by ESA Astronaut Tim Peake)

34. © ESA / NASA (picture taken by ESA Astronaut Tim Peake)
35. © ESA / NASA (picture taken by ESA Astronaut Tim Peake)
36. © ESA / NASA (picture taken by ESA Astronaut Tim Peake)
37. © Getty
38. © ESA / NASA (picture taken by NASA Astronaut Scott Kelly)
39. © ESA / NASA (picture taken by NASA Astronaut Scott Kelly)
40. © ESA / NASA (picture taken by ESA Astronaut Tim Peake)
41. © ESA / NASA (picture taken by NASA Astronaut Tim Kopra)
42. © Getty
43. © Getty

Hardback endpapers © ESA / NASA (pictures taken by ESA Astronaut Tim Peake)